BACKHAULING/ FRONTHAULING FOR FUTURE WIRELESS SYSTEMS

BACKHAULING/ FRONTHAULING FOR FUTURE WIRELESS SYSTEMS

Edited by

Kazi Mohammed Saidul Huq and Jonathan Rodriguez
Instituto de Telecomunicações, Aveiro, Portugal

WILEY

This edition first published 2017
© 2017 John Wiley & Sons, Ltd.

Registered Office
John Wiley & Sons, Ltd, The Atrium, Southern Gate, Chichester, West Sussex, PO19 8SQ, United Kingdom

For details of our global editorial offices, for customer services and for information about how to apply for permission to reuse the copyright material in this book please see our website at www.wiley.com.

Library of Congress Cataloging-in-Publication data applied for

ISBN: 9781119170341

A catalogue record for this book is available from the British Library.

Cover image: Gettyimages/Petrovich9

Set in 11/13pt Times by SPi Global, Pondicherry, India
Printed and bound in Malaysia by Vivar Printing Sdn Bhd

10 9 8 7 6 5 4 3 2 1

Contents

List of Contributors

Jens Bartelt
Technische Universität Dresden, Vodafone Chair MNS, Dresden, Germany

Tsung-Hui Chang
School of Science and Engineering, The Chinese University of Hong Kong, Shenzhen, CUHK (SZ), China

Antonio De Domenico
CEA, LETI, MINATEC, Grenoble, France

Gerhard Fettweis
Technische Universität Dresden, Vodafone Chair MNS, Dresden, Germany

Ummy Habiba
The Department of Electrical and Computer Engineering, University of Manitoba, Canada

Ekram Hossain
The Department of Electrical and Computer Engineering, University of Manitoba, Canada

Aiping Huang
College of Information Science and Electronic Engineering, Zhejiang University, China

Xiaojing Huang
Faculty of Engineering and Information Technology, University of Technology Sydney (UTS), Australia

Kazi Mohammed Saidul Huq
Instituto de Telecomunicações, Aveiro, Portugal

Marios Kountouris
Mathematical and Algorithmic Sciences Lab, France Research Centre, Huawei Technologies, France

Dimitri Ktenas
CEA, LETI, MINATEC, Grenoble, France

Wei-Sheng Lai
Department of Electrical and Computer Engineering, National Chiao Tung
University, Hsinchu, Taiwan

Ta-Sung Lee
Department of Electrical and Computer Engineering, National Chiao Tung
University, Hsinchu, Taiwan

Johannes Lessmann
NEC Laboratories Europe, Heidelberg, Germany

Georgios Mantas
Instituto de Telecomunicações, Aveiro, Portugal

Shahid Mumtaz
Instituto de Telecomunicações, Aveiro, Portugal

Tony Q. S. Quek
Information Systems Technology and Design Pillar, Singapore University of
Technology and Design, Singapore

Jonathan Rodriguez
Instituto de Telecomunicações, Aveiro, Portugal

Peter Rost
Nokia Networks, Munich, Germany

Valentin Savin
CEA, LETI, MINATEC, Grenoble, France

Hangguan Shan
College of Information Science and Electronic Engineering, Zhejiang University,
China

Victor Sucasas
Instituto de Telecomunicações, Aveiro, Portugal

Hina Tabassum
The Department of Electrical and Computer Engineering, University of Manitoba, Canada

Dirk Wübben
University of Bremen, Department of Communications Engineering, Bremen, Germany

Kuan-Hsuan Yeh
ASUSTeK Computer Inc., Taipei, Taiwan

Gongzheng Zhang
College of Information Science and Electronic Engineering, Zhejiang University,
China

Preface

In a mobile communication system, the segment that connects the core to the access networks is termed the 'backhaul'. The edges of any telecommunication network are connected through backhauling. The importance of backhaul research is spurred by the need for increasing data capacity and coverage to cater for the ever-growing population of electronic devices – smartphones, tablets and laptops – which is foreseen to hit unprecedented levels by 2020. The backhaul is anticipated to play a critical role in handling large volumes of traffic, its handling capability driven by stringent demands from both mobile broadband and the introduction of heterogeneous networks (HetNets). Backhaul technology has been extensively investigated for legacy mobile systems, but is still a topic that will dominate the research arena for next generation mobile systems; it is clear that without proper backhauling, the benefits introduced by any new radio access network technologies and protocols would be overshadowed.

Traditionally, the backhaul segment connects the RAN (radio access network) to the rest of the network where the baseband processing takes place at the cell site. However, with the onset of next generation networks, the notion of 'fronthaul access' is also gaining momentum. The future technology roadmap points towards SDN (software-defined networks) and network virtualization as means of effectively sharing resources on demand between different mobile operators, thus taking a step towards reducing the operational and capital expenditure in future networks. Moreover, the baseband processing will be centralized, allowing the operators to completely manage interference through coordinated resource-management strategies. In fact, 3GPP are today visualizing a C-RAN (cloud-RAN) architecture, where the evolved base stations are connected to the C-RAN unit through communication hauls, to what is referred to as the 'fronthaul network'. Traditionally, fibre-optic technology is used to roll out the deployment of base stations; however, this comes along with inherent limitations, including cost and lack of availability at many small sites. This provides the impetus for radio solutions that can handle large volumes of traffic on the fronthaul access, triggering the research community at large to find alternative and advanced solutions that can supersede fibre.

The current work on backhaul and fronthaul technology is fragmented, and still in its infancy. There are still giant steps to be taken towards developing concrete solutions to provide a modern communication haul for next generation networks, which is also commonly referred to as 5G. This book aims to be the first of its kind to hinge together the related discussions on the fronthaul and backhaul access under the umbrella of 5G networks, which we will often refer to as the 'communication haul'. We aim to discuss these pivotal building blocks of the communication infrastructure and provide a view of where it all started, where we are now in terms of LTE/LTE-A networking and the future challenges that lie ahead for 5G. In addition, this book presents a comprehensive analysis of different types of backhaul/fronthaul technologies while introducing innovative protocol architectures.

In the compilation of this book, the editors have drawn on their vast experience in international research and being at the forefront of the communication haul research arena and standardization. This book aims to be the first to talk openly about next generation communication hauls, and will hopefully serve as a useful reference not only for postgraduate students to learn more about this evolving field, but also to stimulate mobile communication researchers towards taking further innovative strides in this field and marking their legacy in the 5G arena.

Kazi Mohammed Saidul Huq
Jonathan Rodríguez
Instituto de Telecomunicações, Aveiro, Portugal

Acknowledgements

This book is the first of its kind tackling the research challenge on the communication haul for legacy and emerging mobile communication networks, and the authors hope that it will serve as a source of inspiration for researchers to drive new breakthroughs on this topic. The inspiration for this book stems from the editors' vast experience at the forefront of European research on backhaul/fronthaul architecture for future wireless systems, including the E-COOP project (UID/EEA/50008/2013), an interdisciplinary research initiative funded by the Instituto de Telecomunicações (Portugal). However, this work would not be complete if it weren't for those who contributed along the way. The editors would first like to thank all the collaborators that have contributed with chapters toward the compilation of this book, providing complementary ideas towards building a complete vision of the communication haul. Moreover, a heartfelt acknowledgement is due to the members of the 4TELL Research Group at the Instituto de Telecomunicações who contributed with useful suggestions and revisions. Furthermore, the editors would like to acknowledge the Fundação para a Ciência e a Tecnologia (FCT-Portugal) for the grant (reference number: SFRH/BPD/110104/2015) that supported this work.

Kazi Mohammed Saidul Huq
Jonathan Rodríguez
Instituto de Telecomunicações, Aveiro, Portugal

1

Introduction: The Communication Haul Challenge

Kazi Mohammed Saidul Huq and Jonathan Rodriguez
Instituto de Telecomunicações, Aveiro, Portugal

1.1 Introduction

Nowadays, the mobile Internet is a pervasive phenomenon that is changing social trends and playing a pivotal role in creating a digital economy. This, in part, is driven by advancements in semiconductor technology, which are enabling faster and more energy-compliant devices, such as smartphones, tablets and sensor devices, among others. However, a truly smart digital world is still in its infancy and the current trends are set to continue, leading to an unprecedented rise in mobile data traffic and intelligent devices. In fact, according to an Ericsson report [1], a typical laptop will generate 11 GB, a tablet 3.1 GB and a smartphone 2 GB per month by the end of 2018. These figures represent the changing communication paradigm, where the end user will not only receive data but generate data; in other words, the end user will become a 'prosumer' running data-hungry applications, for example, high-definition wireless video streaming, machine-to-machine communication, health-monitoring applications and social networking. Therefore, existing technology requires a radical engineering design upgrade in order to compete with ever-growing user expectations and to accommodate the foreseen increase in traffic. The change will be driven by market expectations, and the new technology being considered is fifth generation (5G) communications [2].

Experts anticipate that 5G will deliver and meet the expectations of a new era in wireless connectivity, and will play a key role in enabling this so-called digital world.

In contrast to legacy fourth generation (4G) systems, the widely accepted consensus on the 5G requirement includes [3, 4]:

- Capacity: 1000x increase in area capacity;
- Latency: Less than 1 millisecond (ms) round trip time (RTT) latency;
- Energy: 100x improvement in energy efficiency in terms of Joules/bit;
- Cost: 10–100x reduction in cost of deployment;
- Mobility: Mobility support and always-on connectivity of users that have high throughput requirements.

To achieve these targets, all the key mobile stakeholders, such as operators, vendors and the mobile research community, are contriving to reengineer the mobile architecture in order to support higher-speed data connectivity.

Small-cell technology is an emerging deployment that is providing promising results in terms of delivering fast connectivity due to the small distance between the base station (BS) and the end user, whilst reducing energy consumption. Market use cases of small cells such as the indoor femto cell have already become a success story, so the question is, can we extrapolate the femto cell paradigm to the outdoor world? In fact, current trends are suggesting that this is the way forward, with multi-tier heterogeneous networks being a new design addition to the LTE-Advanced standard [5, 6]. Here, multi-tier radio networks (small-cell tiers) play a pivotal role, coupled with network coexistence approaches to reduce the interference between tiers. Moreover, mobile technology will continue to evolve in this direction with the hyper-dense deployment of small cells providing hotspot islands of high data connectivity coverage zones. This context will ask new questions from the research community in terms of how to tunnel this traffic from the local serving base station towards the core network. Typically, in legacy networks, the segment of the network that interconnects the BS to the RAN (radio access network) to the EPC (evolved packet core) is called the *backhaul*. Fibre optic lines or microwave links have fulfilled this role, with limitations in terms of deployment cost and limited coverage area. However, mobile technology is heading towards an era of virtualization and software-defined networking, where radio resources are allocated from a common pool to different providers, and their management is centralized. This new era is, in fact, reflecting parallels in the cloud computing world, with the onset of cloud services. Emerging mobile networks are heading towards a C-RAN (cloud radio access network) approach [7, 8], where RRUs (remote radio units) and a centralized processing RAN core work in synergy to provide coordinated scheduling, or, in other words, interference management. This paradigm is changing the perception of the communication haul in the network, from backhauling to incorporating both a back and fronthaul segment. In this context, the backhaul dictates how the information is parried from the base stations to the core network, whilst the fronthaul refers to the connectivity segment between the C-RAN core network and the small cell. Figure 1.1 shows definitions of

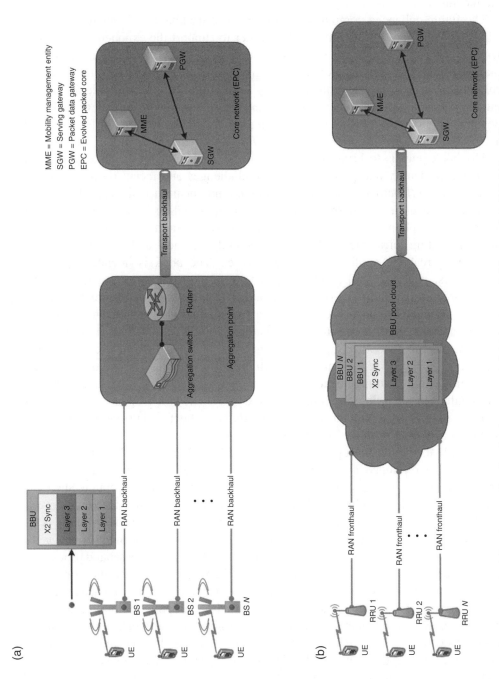

Figure 1.1 Communication haul segments of (a) legacy and (b) emerging C-RAN mobile network

the backhaul and fronthaul segments pertaining to legacy and emerging C-RAN architectures.

The future enhanced communication haul (be it backhaul or fronthaul) for 5G is expected to be deployed around 2020 in order to support the exponential growth in wireless data that is forecast over the next decade. Therefore, there is substantial market interest in the development of ground-breaking backhaul and fronthaul solutions that can not only enhance today's networks, but also provide a coherent interference management approach in emerging technologies such as C-RAN and beyond. This communication haul challenge provided the inspiration for this book and its title: *Backhauling/Fronthauling for Future Wireless Systems*.

The book intends to bring together all mobile stakeholders, from academia and industry, to identify and promote technical challenges and recent results related to smart backhaul/fronthaul research for future communication systems such as 5G. It provides an overview of current approaches to backhauling legacy communication systems and explains the rationale for deploying future smart and efficient backhauling/fronthauling infrastructure from architectural, technical and business points of view using real-life applications and use cases. The book is intended to inspire researchers, operators and manufacturers to render ground-breaking ideas in the newly emerging discipline of smart backhauling/fronthauling over future, ultra-dense wireless systems. Moreover, detailed security challenges are presented to analyse the performance of smart backhauling/fronthauling for future wireless. It is clear that smart backhauling/fronthauling deployment can offer a palette of interesting colours capable of painting new business opportunities for mobile stakeholders for next generation wireless communication systems. This is the first book of its kind on smart backhauling/fronthauling for future wireless systems which updates the research community on the communication haul roadmap, reflecting current and emerging features emanating from the 3GPP group.

To guide the reader through this adventure, the book has the following layout. In Chapter 2, a reference architecture for the future radio communication haul is presented from a 5G perspective. 5G networks are anticipated to obtain Shannon-level and beyond throughput and almost zero latency. However, there are several challenges to solve if 5G is to outperform legacy mobile platforms; one of these is the design of the communication 'haul'. Traditionally, the backhaul segment connects the radio access network (RAN) to the rest of the network where the baseband processing takes place at the cell site. However, in this chapter, we will use the concept of 'fronthaul access,' which is recently gaining significant interest since it has the potential to support remote baseband processing based on adopting a cloud radio access network (C-RAN) architecture that aims to mitigate (or coordinate) interference in operator-deployed infrastructures; this eases significantly the requirements in interference-aware transceivers. To do this, we provide a reference architecture that also includes a network and protocol architecture and proposes a so-called 'cloud resource optimizer'. This integrated solution will be the enabler for

RAN-as-a-Service, not only paving the way for effective radio resource management, but opening up new business opportunities for virtual mobile service providers.

Emerging channel transmission approaches and the possibility of using higher frequency bands, such as massive MIMO and millimetre-wave (mmWave), respectively, are of paramount importance for future wireless systems and for the communication haul. Chapter 3 introduces the fundamentals with regard to massive MIMO and mmWave communication, and their suitability for small-cell backhauling and fronthauling. Furthermore, a performance analysis model for wireless backhauling of small cells with massive MIMO and mmWave communication is outlined. Using this model, some numerical results on the performance of massive-MIMO- and/or mmWave-based wireless backhaul networks are presented.

C-RAN promises considerable benefits compared to decentralized network architectures. Centralizing the baseband processing enables smaller radio access points as well as cooperative signal processing and ease of upgrade and maintenance. Further, by realizing the processing not on dedicated hardware, but on dynamic and flexible general-purpose processors, cloud-based networks enable load balancing between processing elements to enhance energy and cost efficiency. However, centralization also places challenging requirements on the fronthaul network in terms of latency and data rate. This is especially critical if a heterogeneous fronthaul is considered, consisting not only of dedicated fibre but also of, for example, mmWave links. A flexible centralization approach can relax these requirements by adaptively assigning different parts of the processing chain either to the centralized baseband processors or the base stations based on the load situation, user scenario and the availability of the fronthaul links. This not only reduces the requirements in terms of latency and data rate, but also couples the data rate to the actual user traffic. In Chapter 4, a comprehensive overview of different decentralization approaches is given, and we analyse their specific requirements in terms of latency and data rate. Furthermore, we demonstrate the performance of flexible centralization and provide design guidelines on how to set up the fronthaul network to avoid over- or under-dimensioning.

Heterogeneous backhaul deployment using different wired and wireless technologies is a potential solution to meet the demand in small-cell and ultra-dense networks. Therefore, it is of cardinal importance to evaluate and compare the performance characteristics of various backhaul technologies in order to understand their effect on the network aggregate performance and provide guidelines for system design. In Chapter 5, the authors propose relevant backhaul models and study the delay performance of various backhaul technologies with different capabilities and characteristics, including fibre, xDSL, mmWave and sub-6 GHz. Using these models, the authors aim to optimize the base station (BS) association so as to minimize the mean network packet delay in a macro-cell network overlaid with small cells. Furthermore, the authors model and analyse the backhaul deployment cost and show that there exists an optimal gateway density that minimizes the mean backhaul cost per small-cell base station. Numerical results are presented to show the delay performance characteristics of different backhaul solutions. Comparisons between the proposed and traditional BS association policies show the significant effects of backhaul on network performance, which demonstrates the importance of joint system design and optimization for both the radio access and backhaul networks.

The small-cell network (also called a HetNet) has been recognized as a potential solution to offer better service coverage and higher spectral efficiency. However, the dense deployment of small cells could cause inter-cell interference problems and reduce the performance gains of HetNets. Various techniques have been developed in 4G for tackling inter-cell interference. In particular, the inter-cell interference coordination (ICIC) technique can coordinate the data transmission and interference in two neighbouring cells. In Chapter 6, the authors consider a HetNet consisting of macro-cell networks overlaid with small-cell networks that access the same spectrum simultaneously. Here, the HetNet architecture assumes macro cells and small cells interconnected via a high-speed fronthaul/backhaul connection. In particular, due to the mobility of wireless subscribers, the load and data traffic are different in every active macro and small cell. The conventional static enhanced ICIC (eICIC) mechanism cannot ensure that adapting the almost blank subframes (ABS) duty cycle corresponds to the dynamic network condition. Only the dynamic eICIC mechanism is suitable for this non-static network traffic. Therefore, the authors aim to develop a dynamic interference coordination strategy for eICIC for maximizing system utilities under given QoS constraints. In contrast to the traditional eICIC mechanism, the proposed method does not add any backhaul requirements. Computer simulations show that the performance in various scenarios of the dynamic eICIC mechanism with QoS requirements is better than a static eICIC approach and the conventional dynamic eICIC mechanism.

Cell selection for joint optimization considering backhauling technology is needed for future wireless systems. In this regard, Chapter 7 provides a comprehensive analysis for joint optimization considering the backhaul in terms of cell selection. This chapter considers heterogeneous cellular networks, where clusters of small cells are locally deployed to create hotspot regions inside the macro-cell area. Most of the research on this topic has focused on mitigating co-channel interference; however, the wireless backhaul has recently emerged as an urgent challenge to enable ubiquitous broadband wireless services in small cells. In realistic scenarios, the backhaul may limit the amount of signalling that can be exchanged amongst neighbouring cells, which aims to coordinate their operations in real time; furthermore, in highly loaded cells (such as hotspots), the backhaul can limit the data rate experienced by the end users. Here, the authors develop a novel cell-association framework, which aims to balance the users amongst heterogeneous cells to improve the overall radio and backhaul resource usage and increase the system performance. The authors describe the relationship between cell load, resource management and backhaul capacity constraints. Then, the cell-selection problem is expressed as a combinatorial optimization problem and two heuristic algorithms – called *Evolve* and *Relax* – are presented to solve this dilemma. The analysis shows that *Evolve* converges to a near-optimal solution, leading to notable improvements with respect to the classic SINR-based association scheme in terms of throughput and resource utilization efficiency.

High-speed and long-range wireless backhaul is a cost-effective alternative to a fibre network. The ever-increasing demand for high-speed broadband services mandates higher spectral efficiency and wider bandwidth to be adopted in the wireless back-hauls. As wireless mobile networks evolve toward 5G, employing higher-order modulation and performing multiband and multichannel aggregation for wireless backhauling have become industry trends. However, commercially available wireless backhaul systems do not meet the stringent requirements for both high speed and long range at the same time. In Chapter 8, the various system architectures for multiband and multi-channel aggregation are discussed. The challenges for achieving high-speed wireless transmission in multiband and multichannel systems are addressed. These challenges include: how to improve spectrum efficiency and power efficiency; how to prevent inter-channel interference; and how to ensure low latency in order to ensure resilient packet delivery and load balancing.

Despite the significant benefits of C-RAN technology in 5G mobile communication systems, C-RAN technology has to face multiple inherent security challenges associated with virtual systems and cloud computing technology, which may hinder its successful establishment in the market. Thus, it is critical to address these challenges in order for C-RAN technology to reach its full potential and foster the deployment of future 5G mobile communication systems. Therefore, Chapter 9 presents representative examples of possible threats and attacks against the main components in the C-RAN architecture in order to shed light on the security challenges of C-RAN technology and provide a roadmap to overcome the security bottleneck.

In conclusion, we firmly believe this book will serve as a useful reference for early-stage researchers and academics embarking on this radio communication haul odyssey, but beyond that, it targets all major 5G stakeholders who are working at the forefront of this technology to provide inspiration towards rendering ground-breaking ideas in the design of new communication hauls for next-generation systems.

References

[1] Ericsson (2013) *Mobility report*, June.
[2] Andrews, J. G., Buzzi, S., Choi, W., Hanly, S. V., Lozano, A., Soong, A. C. K. and Zhang, J. C. (2014) What Will 5G Be? *IEEE Journal on Selected Areas on Communication*, **32**(6), 1065–1082.
[3] Huawei Technologies Co. (2013) 5G: A technology vision. White paper.
[4] Osseiran, A., Boccardi, F., Braun, V., Kusume, K., Marsch, P., Maternia, M., Queseth, O., Schellmann, M., Schotten, H., Taoka, H., Tullberg, H., Uusitalo, M. A., Timus, B. and Fallgren, M. (2014) Scenarios for 5G mobile and wireless communications: The vision of the METIS project. *IEEE Communications Magazine*, **52**(5), 26–35.

2

A C-RAN Approach for 5G Applications

Kazi Mohammed Saidul Huq, Shahid Mumtaz and Jonathan Rodriguez

Instituto de Telecomunicações, Aveiro, Portugal

2.1 Introduction

Nowadays mobile Internet is a pervasive phenomenon. In the last decade, this phenomenon, along with the market drive for novel software applications spurred by the availability of smartphone handsets, has led to an unprecedented increase in data traffic. Researchers and experts predict that this upward trend will continue as the 5G community envisions new usage scenarios that involve connecting people, machines and applications through a mobile infrastructure. For this reason, the current technology requires a radical change to cater for this new tidal wave of mobile data, which has led us to the fifth generation (5G) communications era [1]. 5G will be expected to deliver a new era of wireless broadband connectivity, shaped by emerging use cases that aim to interconnect devices (the Internet of Things – IoT), enhance quality of experience (QoE) for the end user in terms of traditional mobile connectivity and be the main platform for addressing critical emergency infrastructures. 5G will play a role in the digitalization of Europe, and key targets include: increasing the peak data rate by 100 times, enhancing network capacity by 1000 times, increasing energy efficiency by 10 times and reducing latency by 30 times [2], all of which represent significant and challenging design requirements in contrast to the legacy 4G system. To achieve these targets, mobile stakeholders (such as operators, carriers and manufacturers) are contriving to incorporate macro cells and small cells into the design of the radio access infrastructure. This has forced system designers to reconsider the

Backhauling/Fronthauling for Future Wireless Systems, First Edition.
Edited by Kazi Mohammed Saidul Huq and Jonathan Rodriguez.
© 2017 John Wiley & Sons, Ltd. Published 2017 by John Wiley & Sons, Ltd.

existing backhaul design of legacy 4G radio networks and to consider both a new backhaul and fronthaul design for ultra-dense heterogeneous networks (HetNets).

5G networks are increasingly perceived as carriers to support a fully fledged, data-centric application rather than voice-centric applications. Hence, one of the principal dilemmas operators are coming across nowadays is how to transform the existing backhaul/fronthaul[1] infrastructure into an Internet Protocol (IP)-based backhaul/fronthaul solution for hyper-dense small-cell deployment. With regard to the hauling of data, the continued use of fibre will give rise to the same problems as experienced today, which are mainly economic but also involve restrictions on deployment due to the geographical locations of transceiver cell sites. Millimetre-wave (mmWave) backhaul/fronthaul is an option, but technological and regulatory challenges are yet to be addressed for its successful deployment. Another emerging solution is to exploit the interworking and joint design of open access and backhaul/fronthaul network architecture for hyper-dense small cells based on cloud radio access networks (C-RANs) [3]. This requires smart backhauling/fronthauling solutions that optimize their operations jointly with the access network optimization protocol. The availability, convergence and economics of smart backhauling/fronthauling systems are the most important factors in selecting the appropriate backhaul/fronthaul technologies for multiple radio access technologies (including small cells, relays and distributed antennas) and heterogeneous types of excessive traffic in the future cellular network. However, in this chapter, we will use the concept of 'fronthaul access', which is recently gaining significant interest since it has the potential to support remote baseband processing based on adopting a C-RAN architecture that aims to mitigate (or coordinate) interference in operator-deployed infrastructures; this eases significantly the requirements in interference-aware transceivers. Under the umbrella of a C-RAN scenario, we introduce the notion of a 'cloud resource optimizer', which requires reengineering the medium access control (MAC) to provide a unified solution. The emergence of wireless fronthaul solutions widens the appeal for small-cell deployments, because a fibre-only solution – the technology typically used for fronthaul – is too expensive or just not available at many small-cell sites. Moreover, we will also present a few ideas of potential applications for C-RAN-based mobile systems such as virtualization of device-to-device (D2D) services.

Following the introduction, this chapter is organized as follows. In Section 2.2, we provide a brief overview of different types of backhauling/fronthauling technologies, and in particular, guide the interested reader through the transition from existing to emerging communication haul technologies. In Section 2.3, we present network and protocol architecture for the baseline 3GPP coordinated multi-point (CoMP) system, as a starting point, and then evolve this towards the emerging C-RAN-based architecture in Section 2.4, which is widely seen as the next step on the mobile evolutionary landscape and indeed one step towards the 5G communication platform.

---

[1] The terms backhaul and fronthaul are used interchangeably in this chapter.

Based on this platform, we develop an integrated solution for the cloud resource optimizer, which defines a unified MAC. Section 2.5 takes this design to the next level by using device-to-device (D2D) communication as a use-case application by introducing a new small-cell paradigm based on 'on-demand' virtual small cells for coping with the dynamic variations in mobile traffic throughout the day; which is also an emerging scenario within the context of 5G. Finally, Section 2.6 summarizes and concludes this chapter.

2.2   From Wired to Wireless Backhaul/Fronthaul Technologies

In this section we provide a brief summary of the different kinds of backhaul/fronthaul technologies which are widely accepted and used by operators and service providers. According to [4, 5] hauling technologies are divided into two major categories: wired and wireless. Figure 2.1 shows the classification of backhaul technologies. For example, in the case of the wired backhaul, copper cables are the conventional medium whereas optical fibres are touted as an emerging hauling medium.

In wired backhaul, two types of physical media are widely used: copper cables and optical fibres. Copper cables are the conventional hauling medium between base transceiver stations (BTSs) and the base station controller (BSC) [4]. Currently, copper cables are being replaced by optical fibres due to their higher rates and low latency. Traditional copper-based backhauling is used in digital subscriber line (DSL) access networks [6]. The alternative to copper for mobile backhaul is fibre-based solutions that can provide almost unlimited capacity. The main fibre access options include GPON (gigabit passive optical network), carrier Ethernet and point-to-point (PTP) fibre [7].

There is another type of backhaul: wireless backhaul. This type of communication haul can be distinguished by the different frequency bands. Although the channel traits are different in this type of backhaul owing to different bands, each technology has its own merits and demerits. One very significant similarity amongst these technologies over wired backhaul is fast and relatively cheap deployment. For example, free space optics (FSO) use light to transmit data, but unlike relying on fibre as a transmission medium, free space propagation is applied [8]. FSO links

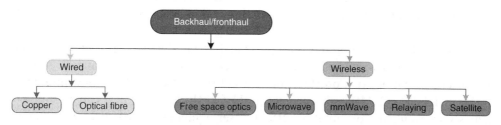

Figure 2.1   Different types of backhaul/fronthaul

also create nearly zero interference between each other; the reason being the narrow beam width. Microwave communication haul technologies utilize different bands of carrier frequencies, ranging from 6 GHz to 42 GHz [5]. Microwave uses licensed spectrum which, in turn, enhances deployment time and cost [9]. Recently, a new paradigm is emerging under the wireless backhaul category: millimetre-wave (mmWave) technology [10]. The explosive developments in circuit technologies have led to mmWave now being considered a viable option, and indeed foreseen as shaping next-generation small-cell wireless backhaul. There are three types of frequency bands available for mmWave – 60, 70/80 and 90 GHz [10]. These high carrier frequencies can enable multi-Gbps data rates [5]. As the 60 GHz band is unlicensed and the higher bands only require an easy and inexpensive licensing process, the links can be deployed much faster and at lower cost [11]. The relay backhaul is another alternative, and is mainly used in the access link. Its inherent advantage is that relays use the same transmission technology and licences as the access link. However, they also have similar shortcomings in terms of range (up to a few kilometres), capacity (a few hundred Mbps) and interference [5]. Satellite backhaul provides an answer for certain terrain where no other backhaul technologies are viable to deploy [4]. In general, T1/E1 is the physical transmission medium over satellite links for cellular backhaul [12].

2.3 Architecture for Coordinated Systems According to Baseline 3GPP

The C-RAN incorporates both a joint signal processing capability and the resource optimization of data belonging to different users which conventional coordinated 3GPP techniques cannot carry out due to high complexity and signal overhead during coordination. Data and signalling are exchanged between different base stations (BSs) through links which are usually capacity limited. This sometimes makes the signalling exchange infeasible. In this section, we describe network and protocol architecture of a coordinated system according to 3GPP.

Figure 2.2 shows the network architecture of a coordinated 3GPP system. This baseline scenario is based on BS cooperation, which recently attracted much interest from the research community. In the 3GPP LTE-Advanced, it is referred to as coordinated multi-point (CoMP) transmission and is being studied actively in LTE release 11 [13].

The inter-BS cooperation has been presented as an effective approach to mitigate inter-cell interference and hence improve cell edge throughput performance. Among the several categories of CoMP technologies [14], we focus only on downlink joint transmission (JT) CoMP in this chapter. In JT CoMP, downlink data can be simultaneously transmitted from multiple BSs to user equipment (UE). It is well known that the cell-edge performance is dramatically improved by JT CoMP. However,

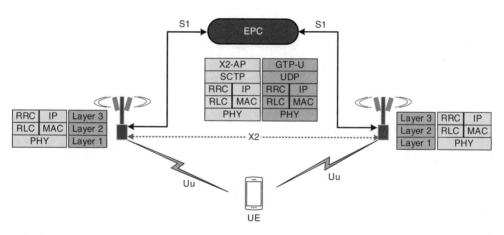

Figure 2.2   Network architecture of baseline 3GPP CoMP system

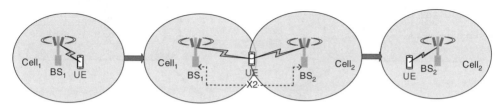

Figure 2.3   Depiction of a JT CoMP use case

the performance of JT CoMP can be degraded in the absence of a high-speed and low-latency backhaul network [15].

This scenario is based on a distributed approach, where each BS has its own layers of LTE protocol stack (i.e., physical (PHY), medium access control (MAC), radio link control (RLC), packet data convergence protocol (PDCP)) and each BS scheduler controls its own UE in the cell. The BSs are connected via an IP-based X2 interface, which acts as an asynchronous communication link for managing JT CoMP operation; this interface is also used for distributing downlink data between BSs. These BSs are attached to the core network via the S1/S5 interface. Moreover, we assume that two BSs are synchronized by a global positioning system (GPS).

To understand the underlying mechanics of CoMP, Figure 2.3 illustrates a JT CoMP use case, where a user migrates between cells in an LTE network. Assume that the UE is located at the cell centre in cell$_1$, initially. The UE is attached to BS$_1$ and receiving a downlink signal from BS$_1$. However, as the UE moves to the cell edge between cell$_1$ and cell$_2$, the UE automatically triggers JT CoMP to improve the performance at the cell edge by receiving a downlink signal from BS$_2$ in addition to BS$_1$. Finally, when the UE moves to cell$_2$, the UE automatically terminates JT CoMP operation and BS$_2$ becomes the communication link.

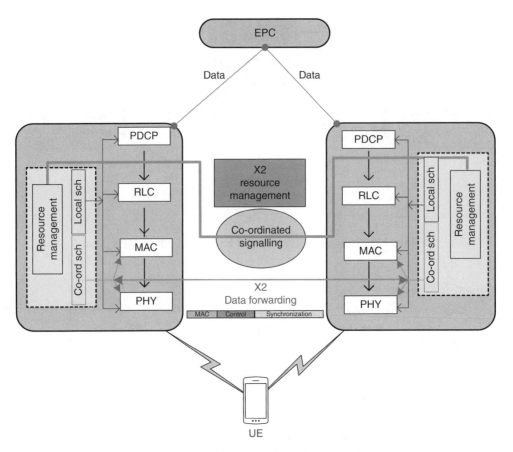

Figure 2.4   Protocol architecture of baseline 3GPP CoMP system

Figure 2.4 shows the protocol architecture to realize the simultaneous transmission scheme based on the LTE standard. The UE reports two kinds of reference signal received power (RSRP) messages to BS_1: $RSRP_1$ from BS_1 and $RSRP_2$ from BS_2. If the difference between $RSRP_1$ and $RSRP_2$ (in dBm) is smaller than the predefined CoMP threshold, then JT CoMP is started; if the difference exceeds the predefined CoMP threshold, then JT CoMP is terminated.

When JT CoMP is triggered, the scheduler in BS_1 will first check with its counter-part in BS_2 to make sure that the radio resources are available for JT CoMP (see the heavy black line in Figure 2.4). During JT CoMP, the downlink data are processed in the following manner (see black arrows). First, PDCP, RLC and MAC are applied to the downlink data in BS_1 and the MAC protocol data unit (MAC-PDU) is created. At the same time, the scheduler in BS_1 provides the joint transmission time as well as control information regarding MCS (modulation and coding scheme), radio resource to be used and antenna mapping for the MAC-PDU. The joint transmission time and the control information are then attached to the MAC-PDU and duplicated; one of them is sent to PHY in BS_1 and other is sent to PHY in BS_2 via the X2 interface. The PHY processing is carried out at both BSs in parallel. Finally, the MAC-PDU is simultaneously transmitted from the two synchronized BSs at the specified joint transmission time.

To transport a MAC-PDU from BS_1 to BS_2, the MAC-PDU is encapsulated by the GTP tunnelling protocol. The joint transmission time and the control information that should be attached to this MAC-PDU are included in a MAC-control element (MAC-CE) in the MAC-PDU.

2.4 Reference Architecture for C-RAN

To overcome the limitations of CoMP, a holistic architectural change is expected via connecting the BSs to central clouds. Unlike the baseline CoMP scenario described in the previous section, in the C-RAN most of the signalling takes place in the cloud and is shared among sites in a pool of virtualized baseband processing units (BBUs). Due to the fact that fewer BBUs are required in the C-RAN compared to the traditional architecture (legacy 3GPP scenario), C-RAN also has the potential to reduce the cost of network operation. This type of network architecture also improves scalability and makes BBU maintenance easier. Different operators can share this cloud BBU pool, which allows some to rent the RAN as a cloud service. Since BBUs from different sites are co-located in one pool, they can communicate with lower delays. This brings to the forefront many other advantages, since existing mechanisms introduced in LTE-A to increase spectral efficiency, interference management and throughput, such as enhanced inter-cell interference coordination (eICIC) and CoMP, are greatly facilitated here.

2.4.1 System Architecture for Fronthaul-based C-RAN

Emerging scenarios in cell deployment are heading towards the notion of cloud radio. In this section we provide the reference system model for the C-RAN scenario with the description of its components. C-RAN is a novel mobile technology that separates baseband processing units (BBUs) from radio front-ends such as remote radio units (RRUs). In this technology, BBUs of several BSs are positioned in a central entity to form a BBU pool where the radio front-ends of those BSs are deployed at the cell sites [16–18]. Therefore, this new framework unfolds a new paradigm for algorithms/techniques that require centralized and cooperative processing. However, the deployment of this new technology faces several potential research challenges, which include latency, efficient fronthaul design and radio resource management for a converged network.

Fronthaul enables a C-RAN architecture in which all the BBUs are placed at a distance from the cell site. The fronthaul transports the unprocessed RF signal from the antennas to the remote BBUs. While the fronthaul requires higher bandwidth, lower latency and more accurate synchronization than the backhaul, it does support more efficient use of RAN resources; when coupled with legacy interference and mobility management tools, this can significantly minimize interference in the structured part of the network, including multi-tier cell interference.

The general system model of the fronthaul-based C-RAN scenario is illustrated in Figure 2.5, and consists of three main components [18], namely: (i) a centralized BBU pool, (ii) remote radio units (RRUs) with antennas and (iii) a transport link, that is a fronthaul network which connects the RRUs to the BBU pool. The RRU provides the interface to the fibre as well as performing digital processing, digital-to-analogue conversion, analogue-to-digital conversion, power amplification and filtering [16]. The distance between the RRU and the BBU can be extended up to 40 km, where the ceiling range emanates from the processing and propagation delay. Optical fibre, mmWave and microwave connections can be used. In the downlink, the RRUs transmit the RF signals to the UEs, or in the uplink the RRUs carry the baseband signals from the UEs to the BBU pool for further processing. The BBU pool is composed of BBUs which operate as virtual base stations to process baseband signals and optimize the network resource allocation for one RRU or a set of RRUs. The fronthaul links can constitute different technologies, namely wired (fiber → ideal) and wireless (mmWave → non-ideal). One can easily add or update any number of BBUs in this cloud depending on the needs and cell planning of the network operator. This C-RAN-based architecture is also more energy efficient than the CoMP-based scenario due to reduced power consumption at the cell sites. In the C-RAN network architecture, no additional power is needed in cell sites other than for RRU operation.

By enabling joint processing in the cloud, key research challenges emerge related to joint provisioning of resources between the different BBUs. This leads us to the design of a so-called 'cloud resource optimizer'.

2.4.2   Cloud Resource Optimizer

In this section we present the proposed cloud resource optimizer for the C-RAN. Interconnections and functions split between BBUs and RRUs are depicted in Figure 2.6. Unlike a CoMP resource management module, where all the resource management entities are separated for different BSs, this resource optimizer unifies all the resource management operation including allocation, interference management and signalling for different BBUs in the cloud pool. Inside this cloud resource optimizer, the PHYs from different RRUs are merged into one common MAC, control (Ctrl) and Synchronization (Sync) entity. This operation prompts us to develop a new MAC approach for this cloud-based system. The MAC works as an enabler between different types of radio access technologies (RAT) such as LTE (IMT technology) and WiFi (non-IMT technology).

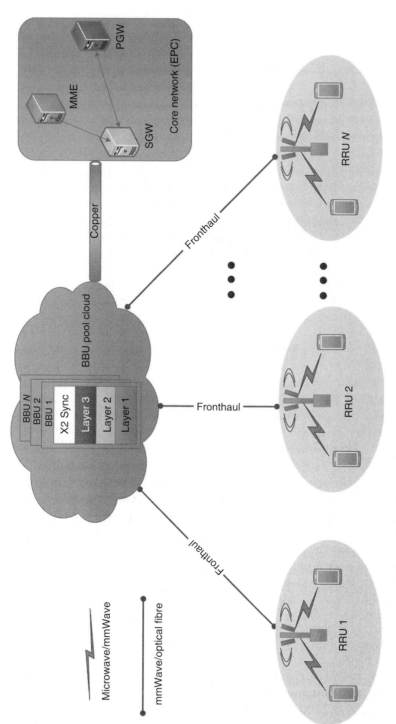

Figure 2.5 Operator's perspective on the fronthaul-based C-RAN scenario

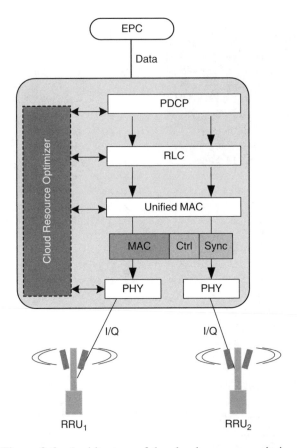

Figure 2.6   Architecture of the cloud resource optimizer

We consider a novel, unified MAC frame for our C-RAN scenario in Figure 2.7, unlike in legacy CoMP where each RAN has its own MAC. The shift in engineering design to introduce the presence of a global MAC entity will not only improve the efficiency (both spectrum and energy) of the system, but take a step towards reducing the overall interference in the network. This unified MAC will be a modified version of an existing LTE MAC frame described in [19].

As can be seen in Figure 2.7, there are several MAC-CEs in both the downlink and uplink MAC. Following Table 1 and Table 2 from the 36.321 standard [19] (shown here in Tables 2.1 and 2.2), we can see the logical channel ID (LCID) types of MAC header. The parts indicated by the bold rectangle emphasize the LCID values for the various MAC-CEs.

We define a new MAC-CE for this purpose. We use the reserved element field for specifying the unified frame, and this is indexed in the MAC-PDU sub-header by an LCID value equal to 11001 in the uplink. The new element is called a unified frame and is appended to the existing LCID values, such as the common control channel (CCCH), cell radio network temporary identifier (C-RNTI) and the padding.

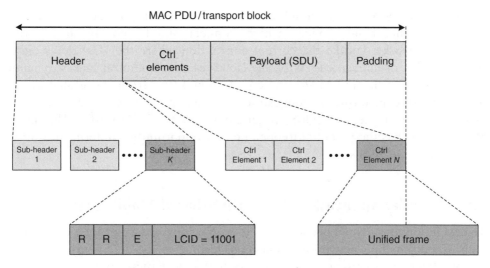

Figure 2.7   Unified MAC frame for C-RAN

Table 2.1   Values of LCID for DL-SCH (adapted from [19])

Index	LCID values
00000	CCCH
00001–01010	Identity of the logical channel
01011–11011	Reserved
11100	UE Contention Resolution Identity
11101	Timing Advance Command
11110	DRX (Dynamic Reception) Command
11111	Padding

Table 2.2   Values of LCID for UL-SCH (adapted from [19])

Index	LCID values
00000	CCCH
00001–01010	Identity of the logical channel
01011–11001	Reserved
11010	Power Headroom Report
11011	C-RNTI
11100	Truncated BSR (Buffer Status Report)
11101	Short BSR
11110	Long BSR
11111	Padding

This unified MAC in the cloud resource optimizer can provide differentiated services to dual-band devices which connect simultaneously to different kinds of networks. Unlike the current convention where the UE selects either licensed→LTE or unlicensed→WiFi, the cloud resource optimizer makes dynamic decisions in the unified MAC framework, benefiting from the global knowledge available in terms of congestion level on the different radio networks and the quality of service (QoS) requirements of the UE's traffic. This novel MAC scheme has the potential to open up new opportunities in terms of traffic-orientated applications.

2.5 Potential Applications for C-RAN-based Mobile Systems

The evolution towards 5G is considered to be the convergence of Internet services with existing mobile networking standards, leading to the commonly used term 'mobile Internet' over small cell, with very high connectivity speeds. In addition, green communications seem to play a pivotal role in this evolutionary path, with key mobile stakeholders driving momentum towards a greener society through cost-effective design approaches. In fact, it is becoming increasingly clear from emerging services and technological trends that energy and cost per bit reduction, service ubiquity and high-speed connectivity are becoming desirable traits for next-generation networks. Providing a step towards this vision, small cells are envisaged as the vehicle for ubiquitous 5G services providing cost-effective, high-speed communications.

These small cells are set up on demand and constitute a novel paradigm toward a 5G wireless system in two ways:

- A wireless network of cooperative small cells (CSC);
- Virtual small cells (VSC).

These novel, on-demand small cells (CSC, VSC) have the ability to cope with today's wireless traffic dynamics and adjust their RF parameters accordingly. Moreover, these on-demand small cells can be used in various applications and scenarios en route to the accomplishment of the 5G goal. Among others, one of these is D2D-based C-RAN, which will be discussed in the next section.

2.5.1 Virtualization of D2D Services

D2D is widely considered an effective and efficient candidate approach for very low latency communications in 5G [20], as well as being the enabler for improving spectrum efficiency. In fact, by reusing the spectrum, two D2D users can form a direct data link without exploiting explicitly the communication infrastructure

(BS and core networks). D2D communications will also be spurred on from the application perspective, as D2D is considered an ideal deployment for proximity-based services, an example being social networking.

However, coexistence approaches are required in order for D2D users not to interfere with macro-cell users, but at the same time to exploit spectrum appropriately in an era where spectral resources are at a premium. In this context, we propose an integrated solution where we combine technology paradigms such as C-RAN and virtual small cells in synergy to provide an enabler for effective D2D-based communications [21]. This novel architecture has the potential to solve most of the challenges related to emerging 5G systems (capacity, latency, energy efficiency, CAPEX/OPEX and mobility). Moreover, these D2D networks will be created on demand, for example, if there are certain users at the cell edge with physically low battery levels, then a user with a high battery level will become the cluster head and other users with low battery levels will communicate in a D2D manner with the cluster head, while the cluster head communicates directly with the C-RAN.

In this architecture, we first split the control/data plane where the RRU provides the signalling service for the whole coverage area and exploit these virtual small cells for delivering data services for high-rate transmission, complemented by a light control overhead and a selection option on the most appropriate air interface (mmWave could be the best option), which is illustrated in Figure 2.8.

2.5.2 Numerical Analysis

In order to analyse the performance of D2D-based C-RAN efficiently, we have enhanced an existing system-level simulator (SLS) with a centralized cloud entity to control all the baseband processing, an RRU which acts as an antenna and a D2D user pair. Moreover, we have also enhanced the following key performance indicators (KPIs) to evaluate the performance of the proposed system.

- **Throughput (with ideal fronthaul):** The average throughput per cell is defined as the sum of the total amount of bits being successfully received by all active users in the system divided by the product of the number of cells being simulated in the system and the total amount of time spent in the transmission of these packets (the simulation time for LTE is TTI = 1 ms).
- **Throughput (with non-ideal fronthaul):** The average throughput per cell is defined as the sum of the total amount of bits being successfully received by all active users in the system divided by the product of the number of cells being simulated in the system, the total amount of time spent in the transmission of these packets (the simulation time for LTE is TTI = 1 ms) and the delay of the fronthaul link (10 ms).

Table 2.3 shows the full list of simulation parameters.

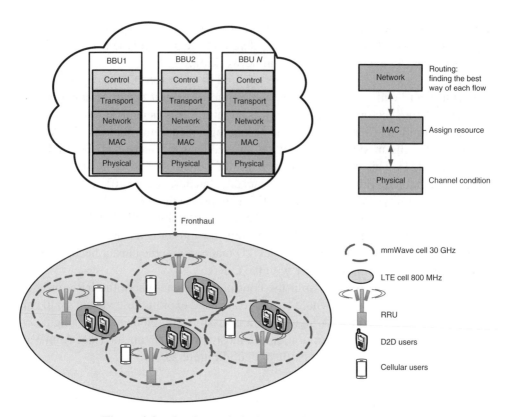

Figure 2.8   On-demand, device-centric advanced C-RAN

Figure 2.9 shows the average throughput of the system with and without ideal fronthaul. For this simulation, we consider 20 D2D pairs and 20 cellular users (CUs). We also assume a fixed resource allocation between CUs and D2D users. There are 100 resource blocks (RBs) in the LTE 20 MHz band, which are divided equally among CUs and D2D users (50 RBs each). Each of these RBs is then assigned to its corresponding users via a proportional fairness (PF) scheduler. CUs on 50 RBs communicate using a cellular link (UE1↔RRU↔UE2), while D2D users use a direct link (UE1↔UE2). When only CUs with ideal fronthaul are deployed, the average throughput of the system is around 10 Mbit/s, but when this scenario is enhanced with D2D, the average throughput rises to 20 Mbit/s. This increase in throughput is due to the inclusion in the C-RAN network of D2D, which, thanks to its direct communication capability, enhances the average throughput of the system. Moreover, if resource-allocation schemes between CUs and D2D users with some interference-cancellation mechanism are considered, the average throughput of the system increases even further.

Table 2.3 Simulation parameters

Name		Parameter
System		LTE-A, 20 MHz, 2.6 GHz
Resource block (RB)		100
Duplexing method		Cellular: FDD (downlink)
		D2D: FDD (uplink using TDD timeslot)
Mode selection		Shortest distance (cellular or D2D)
Resource allocation		Fixed allocation
Channel estimation		Perfect
Channel models	Between D2D	$40\log10d[m]+30+30\log10(f\,[\text{Mhz}]+49)$
	RRU→ D2D	$36.7\log10d[m]+40.9+26\log10(f\,[\text{Mhz}]/5)+\alpha_{shadowing}$
	RRU→ CU	$36.7\log10d[m]+40.9+26\log10(f\,[\text{Mhz}]/5)+\alpha_{shadowing}$
Retransmission		HARQ
Scheduler of eNB		Proportional fairness (PF)
Power control		Adaptive power
Traffic		Full buffer
Fronthaul		Ideal (no delay)/non-ideal (10 ms delay)
Maximum transmit power		RRU = 30 dBm,
		Cellular Tx_Power = 24 dBm
		D2D Tx_Power = 9 dBm
Noise figure		5 dBm for base station/9 dBm for D2D receiver
Thermal noise density		−174 dBm/Hz
User speed		Static

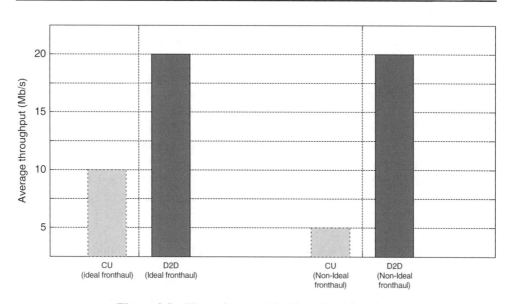

Figure 2.9 Throughput vs. ideal/non-ideal fronthaul

When simulations with a non-ideal fronthaul are run, a delay of 10 ms is experienced, as shown in Figure 2.9, and the throughput of CUs drops to around 4 Mbit/s. This is due to the fronthaul delay: the greater the delay, the lower the throughput. But for the D2D case, throughput remains the same, because in D2D, data are transferred directly between devices and therefore there is a sort of 'zero delay' [22], but this is still under the control of the C-RAN, which is a benefit in terms of mobility and handover.

Virtual small cells not only need to be ubiquitous and cost-effective, but have to deliver emerging services in a secure fashion in an era where applications will handle 'extremely confidential data' and money transactions. Therefore, 5G networks must deliver a framework with a palette of security tools that are the enablers for a cross-system, end-to-end secure link which is fast and lightweight in nature. Virtual small cells also open up the possibility for network operators to invest in network-sharing scenarios whereby operators can accommodate the foreseen increase in traffic whilst reducing their investment in new infrastructure, and beyond that significantly reduce their energy bill. These enhancements are currently addressed by 3GPP; however, this raises new research challenges when applied to the virtual small environment.

Table 2.4 shows the comparison of D2D-based C-RAN with existing LTE-Advanced (LTE-A) technologies like CoMP, taking into consideration the main architectural blocks of the communication network, like the evolved node B (eNB) (in LTE, the BS is called the eNB).

2.6 Conclusion

To take a step towards the 5G vision, this chapter has described a C-RAN reference system architecture as the fundamental building block that has the potential to evolve and to anticipate disruptive changes in users' demands and small-cell deployment. The first part of this chapter introduces the C-RAN architecture, which exploits RRU technology connected to the core network using backhaul technology based on fibre-optic links. C-RAN was engineered so as to mitigate one of the key disabling features affecting mobile communications since their inception: user interference. In C-RAN, the key is to harness all the baseband processing from all users within a common unit, thus providing the operator complete control over the network and the ability to coordinate signal transmissions, providing a significant step towards mitigating interference in the network and alleviating interference-aware transceivers. However, the key aim of the chapter was to go beyond C-RAN and examine the fronthaul component. In particular, we use the C-RAN approach as a fundamental building block and build on this to provide a more flexible platform that is able to support emerging use cases under the 5G umbrella. In this context, we introduce the notion of a cloud resource optimizer which works in synergy with a unified MAC, allocating virtual radio resources on demand to support various applications, and potentially acting as the lynchpin to

Table 2.4 Comparison of technologies

Characteristic	CoMP	C-RAN	D2D [23]	D2D:C-RAN
Standardization	3GPP release 11–12	IEEE	3GPP release 11–12	3GPP and IEEE
Frequency band	Licenced bands	Licenced bands	Licenced bands	Licenced/Unlicenced bands
Max transmission distance	500 m	100 m/1000 m[1]	20 m	100 m/1000 m
Capacity[2]	Good (500–600 Mbps)	Good (500–600 Mbps)	Very good (1 Gbps)	Excellent (2 Gbps)
Latency	>1 ms	>1 ms	'Zero latency'	'Zero latency'
Uniformity of service provision	No	No	Yes	Yes
Application	Improved capacity at cell edge. Longer battery life of hand-sets. Complete operator control.	Improved capacity and coverage at cell centre and edge. More energy efficient than CoMP.	Improved capacity and coverage at cell centre and edge. More energy efficient than CoMP. Relaying and C-RAN in cellular networks, safety in public services, sharing in context-based applications.	Combination of C-RAN and D2D.
Infrastructure	Within licensed bands, a central controlling unit such as eNB is used for transferring data.	Within licensed bands, a radio controlling cloud is used for transferring data.	Transferring of data between users occurs directly, be it licensed or unlicensed bands.	Transferring of data between users occurs in licensed bands directly. It is managed by a cloud central controller.

(continued)

Table 2.4 (continued)

Characteristic	CoMP	C-RAN	D2D [23]	D2D:C-RAN
Expenses	**CAPEX:** Subsidized hardware. Commissioning new cell sites and towers. **OPEX:** Communication haul, i.e., fronthaul. Leasing and electric power consumption of the cell site. Operational cost of eNBs.	**CAPEX:** Subsidized C-RAN (BBU) hardware. Installing new RRUs. **OPEX:** Communication haul, i.e., fronthaul. Leasing and electric power consumption of the cell site. Operational cost of RRUs.	**CAPEX:** No expenses since users are using their devices with all types of technologies such as WiFi, LTE and D2D. **OPEX:** Usage of battery of the devices.	Combination of C-RAN and D2D.

[1]100 m is for mmWave at 60 GHz and 1000 m is for fibre.

[2]These values are calculated under ideal channel and traffic conditions and with 2x2 MIMO using 20 Mhz of bandwidth.

support the design of new 5G protocols/algorithms. In particular, we show how this platform can not only support the emerging D2D paradigm, but is also able to support the deployment of small-cell technology which is seen as pivotal to the 5G story.

References

[1] Andrews, J. G., Buzzi, S., Choi, W., Hanly, S. V., Lozano, A., Soong, A. C. K. and Zhang, J. C. (2014) What Will 5G Be? *IEEE Journal on Selected Areas in Communications*, **32**(6), 1065–1082.

[2] Osseiran, A., Boccardi, F., Braun, V., Kusume, K., Marsch, P., Maternia, M., Queseth, O., Schellmann, M., Schotten, H., Taoka, H., Tullberg, H., Uusitalo, M. A., Timus, B. and Fallgren, M. (2014) Scenarios for 5G mobile and wireless communications: the vision of the METIS project. *IEEE Communications Magazine*, **52**(5), 26–35.

[3] China Mobile Research Institute (2011) 'C-RAN: The Road Towards Green RAN.' Technical report, April. Available at: http://labs.chinamobile.com/cran/wp-content/uploads/CRAN_white_paper_v2_5_EN.pdf.

[4] Tipmongkolsilp, O., Zaghloul, S. and Jukan, A. (2011) The Evolution of Cellular Backhaul Technologies: Current Issues and Future Trends. *IEEE Communications Surveys Tutorials*, **13**(1), 97–113.

[5] Bartelt, J., Fettweis, G., Wubben, D., Boldi, M. and Melis, B. (2013) 'Heterogeneous Backhaul for Cloud-Based Mobile Networks.' Paper presented at the Vehicular Technology Conference (VTC Fall), pp. 1–5.

[6] Eriksson, P. and Odenhammar, B. (2006) 'VDSL2: Next important broadband technology,' Ericsson Review No. 1.

[7] Orphanoudakis, T., Kosmatos, E., Angelopoulos, J. and Stavdas, A. (2013) Exploiting PONs for mobile backhaul. *IEEE Communications Magazine*, **51**(2), S27–S34.

[8] LightPointe Communications Inc. (2009) 'Understanding the performance of free space optics.' White paper.

[9] Giesken, K. (2002) Application of wireless technology in the mobile backhaul network. *Bechtel Telecommunications Technical Journal*, **1**(1), 62–70.

[10] Rappaport, T. S., Sun, S., Mayzus, R., Zhao, H., Azar, Y., Wang, K., Wong, G. N., Schulz, J. K., Samimi, M. and Gutierrez, F. (2013) Millimeter Wave Mobile Communications for 5G Cellular: It Will Work! *IEEE Access*, **1**, 335–349.

[11] Rangan, S., Rappaport, T. S. and Erkip, E. (2014) Millimeter-Wave Cellular Wireless Networks: Potentials and Challenges. *Proceedings of the IEEE*, **102**(3), 366–385.

[12] Owens, J. (2002) Satellite Backhaul Viability. *Bechtel Telecommunications Technical Journal*, **1**(1), 58–61.

[13] RP-111117 Work Item Description, 'Coordinated Multi-Point Operation for LTE,' Samsung, 3GPP TSG RAN meeting #53, Fukuoka, Japan, September 13–16, 2011.

[14] Zhang, L., Nagai, Y., Okamawari, T. and Fujii, T. (2013) 'Field Experiment of Network Control Architecture for CoMP JT in LTE-Advanced over Asynchronous X2 Interface.' Paper presented at the Vehicular Technology Conference (VTC Spring), 2nd–5th June, pp. 1, 5.

[15] Okamawari, T., Zhang, L., Nagate, A., Hayashi, H. and Fujii, T. (2011) 'Design of Control Architecture for Downlink CoMP Joint Transmission with Inter-BS Coordination in Next Generation Cellular Systems.' Paper presented at the Vehicular Technology Conference (VTC Fall), 5th–8th September, pp. 1–5.

[16] Checko, A., Christiansen, H. L., Yan, Y., Scolari, L., Kardaras, G., Berger, M. S. and Dittmann, L. (2015) Cloud RAN for Mobile Networks – A Technology Overview. *IEEE Communications Surveys Tutorials*, **17**(1), 405–426.

[17] Beyene, Y. D., Jantti, R. and Ruttik, K. (2014) Cloud-RAN Architecture for Indoor DAS. *IEEE Access*, **2**, 1205–1212.

[18] Wang, R., Hu, H. and Yang, X. (2014) Potentials and Challenges of C-RAN Supporting Multi-RATs Toward 5G Mobile Networks. *IEEE Access*, **2**, 1187–1195.

[19] http://www.etsi.org/deliver/etsi_TS/136300_136399/136321/09.00.00_60/ts_136321v090000p.pdf.

[20] Mumtaz, S., Huq, K. M. S. and Rodriguez, J. (2014) Direct mobile-to-mobile communication: Paradigm for 5G. *IEEE Wireless Communications*, **21**(5), 14–23.

[21] Huq, K. M. S., Mumtaz, S., Rodriguez, J., Marques, P., Okyere, B. and Frascolla, V. (2016) Enhanced C-RAN using D2D Network. *IEEE Wireless Communications*, submitted January, 2016 (under review).

[22] Huawei Technologies Co. (2013) '5G: A technology vision.' White paper.

[23] Feng, D., Lu, L., Yuan-Wu, Y., Li, G., Li, S. and Feng, G. (2014) Device-to-device communications in cellular networks. *IEEE Communications Magazine*, **52**(4), 49–55.

3

Backhauling 5G Small Cells with Massive-MIMO-Enabled mmWave Communication

Ummy Habiba, Hina Tabassum and Ekram Hossain
Department of Electrical and Computer Engineering, University of Manitoba, Canada

3.1 Introduction

The exponential growth in the number of mobile subscribers and their corresponding wireless data traffic has demanded the emergence of fifth generation (5G) cellular networks. One of the key requirements for 5G is to increase the data rate radically, to approximately 1000 times that of the current 4G technology. In this regard, network densification is a straightforward way to increase the data rates. The idea of network densification is to increase the density of small base stations (SBSs) in order to enhance the number of users supported in a geographic region [1]. In principle, the density of SBSs can be increased indefinitely until there is only one user supported per SBS with its transmission and backhaul connections [1]. This extreme densification raises a variety of challenges that include determining the appropriate cell associations, managing inter-tier and intra-tier interference and providing high-capacity backhaul connectivity to SBSs at the same time.

Provisioning of cost-effective and scalable backhaul solutions for 5G ultra-dense networks is challenging due to the need for ample resources for the backhaul transmissions to/from a massive number of SBSs. Consequently, it is crucial to consider both wired and wireless options for backhauling infrastructure of 5G cellular networks depending on the location of the small cells and quality-of-service (QoS)

Backhauling/Fronthauling for Future Wireless Systems, First Edition.
Edited by Kazi Mohammed Saidul Huq and Jonathan Rodriguez.
© 2017 John Wiley & Sons, Ltd. Published 2017 by John Wiley & Sons, Ltd.

requirements of the users. Even though the existing wired backhaul ensures high reliability, it may not be an economical and ascendable solution for densely deployed SBSs. On the other hand, wireless backhauling solutions are cost-efficient as well as easily scalable to provide connectivity to small cells. Reasons include the possibility of frequency reuse, easy operation and management and increased flexibility of backhaul resource allocation.

There exist several wireless backhauling solutions for small cells, such as TVWS (< 800 MHz), sub-6 GHz (licensed and unlicensed), microwave spectrum between 6 GHz and 60 GHz and FSO (free space optical) spectrum within the laser spectrum. Moreover, the spectrum available between 30 and 300 GHz can offer 200 times greater resources than the resources used for current cellular communications [2]. This frequency band with the wavelengths ranging from 1 to 10 mm is referred to as millimetre-wave (mmWave) spectrum. Among various wireless backhauling options, mmWave spectrum between 60 and 90 GHz is considered suitable for 5G networks due to its small wavelength which allows the use of a large number of antenna elements in a compact form, line-of-sight (LOS) propagation with larger channel bandwidth and ample unlicensed spectrum.

Nonetheless, mmWave spectrum has a number of propagation challenges due to its susceptibility to shadowing, rain attenuation and molecular absorption. The mmWave signals cannot penetrate building walls and other blockages. Therefore, indoor and outdoor users need to be served by separate BSs operating in mmWave frequencies with large directivity gains. Consequently, highly directive antennas are essential for mmWave BSs to generate narrow beams that suppress the impact of environmental attenuation and interference arriving from neighbouring cells.

In this context, large-scale antenna systems (often referred to as massive MIMO) can be deployed at the transmitter and receiver to enhance the directivity. The massive MIMO technology is well suited to the short-wavelength mmWave signals, as a large number of antenna elements can be installed within a small area of mmWave transceiver nodes, enabling highly directional beams with high capacity. Massive MIMO technology coupled with efficient beamforming strategies diminishes the uncorrelated noise and short-term fading effects irrespective of the number of users or BSs in the network [3]. However, the performance of beamforming remains limited by the channel estimation and acquisition techniques.

The rest of the chapter is organized as follows. In Section 3.2, we first compare different wireless backhauling solutions that currently exist. Section 3.3 describes the key features of mmWave propagation and massive MIMO technology. In Section 3.4, we review the state-of-the-art in designing backhaul systems using mmWave spectrum and discuss different LOS and NLOS topologies for mmWave backhauling. In this section, we also outline the design and implementation issues for massive-MIMO-enabled mmWave backhauling systems. Next, in Section 3.5, we model a massive-MIMO-enabled mmWave backhauling system and consider maximizing the network rate through an efficient stable-matching-based user association scheme. Numerical results are also presented in Section 3.5 before the chapter is concluded in Section 3.6.

3.2   Existing Wireless Backhauling Solutions for 5G Small Cells

The wireless backhaul connectivity of small cells is a function of the location and density of the SBSs, line-of-sight (LOS) conditions between the wireless backhaul hubs and SBSs, data rate requirements and additional costs for spectrum licensing. Appropriate backhauling spectrum thus needs to be chosen as per the network requirements. To this end, in this section, based on the studies in [4–6], we detail the features of different spectrum options for wireless backhauling.

- *TVWS (600–800 MHz):* The unused TV spectrum, known as TV white spaces (TVWS), can be a suitable option for sparsely populated areas where communication is limited to voice, video and real-time gaming. This unlicensed spectrum can be accessed opportunistically to backhaul small cells in a cognitive manner. TVWS allows NLOS backhaul connectivity over a wider area (1–5 km) without any constraint for antenna alignment. Due to opportunistic access, there is a risk of unavailability of the spectrum. In addition, the backhaul connectivity must ensure that it does not impose any interference on primary TV transmission. The data rates provided by TVWS-based backhauls may not be sufficient for backhauling 5G networks.
- *Sub-6 GHz:* The licensed sub-6 GHz spectrum (800 MHz – 6 GHz) is suitable for backhauling SBSs in rural as well as urban areas. Without any additional hardware or antenna alignment requirement, this spectrum allows NLOS communication over a wider distance. In urban areas, it can provide backhaul coverage from 1.5 to 2.5 km whereas it can support backhaul links up to 10 km in rural areas. However, the sub-6 GHz spectrum is costly and provides narrow channel bandwidth which limits the capacity. In dense areas, sub-6 GHz spectrum is already in use; therefore, it can suffer from severe interference. Sub-6 GHz spectrum is thus not recommended for backhauling 5G small cells in dense urban areas.
- *Microwave (6–60 GHz):* The microwave spectrum is a common choice for backhauling SBSs in urban and rural areas as it can provide 1 Gbps+ data rates for real-time and non-real-time services. However, the spectrum is costly and it requires antenna alignment for LOS communication to achieve high directivity gain. The size of the antennas designed for these frequency bands is relatively large, which might not be feasible for SBSs. The microwave signals can provide backhaul coverage up to 4 km. Therefore, microwave backhauls may be beneficial from the perspective of data rate but may not be cost-effective.
- *Unlicensed millimetre wave (60–90 GHz):* The unlicensed mmWave spectrum is another promising candidate for backhauling 5G small cells in dense urban areas. Since mmWave signals suffer from high propagation loss, a coverage distance

beyond 1 km may not be possible. However, this may work for densely deployed small cells because of the reduced interference from adjacent links and, in turn, increased capability of frequency reuse. These extremely high frequency signals enable the installation of a large number of antenna elements within a small area which enable high antenna array gain. The wider channel bandwidth offered by mmWave spectrum and high directivity can provide multi-Gbps data rates. Although mmWave spectrum requires antenna alignment and LOS connectivity, there is no additional spectrum cost.

• *Free space optical (FSO) technology:* FSO can be an alternative solution to traditional radio frequency (RF) spectrum. The FSO link (laser beam) between laser photo-detector transceivers is not susceptible to electromagnetic interference. FSO links use wavelengths in the range of micrometre and are capable of providing data rates in the order of 10 Gbps over 1 km [7]. There is no licensing cost for FSO transmission. However, there will be a hardware cost which is comparatively lower than optical fibre links. For FSO links, LOS is required and laser beams are sensitive to weather conditions such as rain, snow and fog. To minimize the effects of atmospheric atten-uation, several techniques can be used that include aperture averaging, wave division multiplexing (WDM), larger receive aperture, fine pointing mirrors (FPM), etc.

Among the existing wireless backhaul solutions, mmWave spectrum has the potential to meet the requirements of 5G networks as it can provide wider channel bandwidth with highly directive narrow beams and, in turn, low interference. The small wavelength of mmWave also has the flexibility to incorporate the massive MIMO technology and, in turn, support a large number of users with a high data rate. As such, in dense urban areas, massive-MIMO-enabled mmWave links are highly suitable for backhauling 5G small cells. In the following section we discuss the key features of mmWave and massive MIMO technologies.

3.3 Fundamentals of mmWave and Massive MIMO Technologies

3.3.1 MmWave Communication

As mentioned in the previous section, the capability of providing larger spectrum with reduced interference and a high data rate makes mmWave spectrum attractive for backhauling ultra-dense cellular networks. Nonetheless, the challenging features of mmWave propagation include the following:

• *Atmospheric attenuation*: The extremely high frequencies in the mmWave spec-trum are vulnerable to high propagation losses due to rain attenuation, oxygen or other molecular absorption. However, these environmental factors may not cause significant propagation loss when considered for short-range communications, for example, ultra-dense small-cell networks. The results in [8] imply a minimal impact of rain attenuation on mmWave frequencies. In times of heavy rainfall, the rain attenuation for 28 GHz is only 1.4 dB over 200 m distance [9]. The attenuation due to atmospheric absorption is also very low for mmWave frequencies, particularly for 28 GHz and 73 GHz [9]. In the case of 60 GHz, oxygen absorption may cause attenuation from 15 to 30 dB/km [10].

- *Penetration losses*: The mmWave frequencies experience high penetration loss on the exterior surfaces of urban buildings whereas the penetration loss for indoor materials is relatively low [9]. Thus, the outdoor BSs in the mmWave network can hardly serve indoor users.
- *Reflection factor*: Large numbers of obstructions may affect the reflected multi-paths of mmWave signals with large delays for both LOS and NLOS situations. Even then, strong signals can be received within 200 m distance in a highly NLOS environment [9]. Path-loss exponents for the mmWave frequencies are within the following range: 3.2–4.58 for NLOS and 1.68–2.3 for LOS environments [9, 11].

For a viable implementation of mmWave backhauling, the existing system designs need to be improved. For instance, the high-gain and electrically steerable antennas at the user and BSs with appropriate beamforming can produce highly directive beams and can combat large-scale and small-scale fading in the channel. The directivity gain can be improved further by using spatial multiplexing. To achieve high antenna gain in dense small-cell networks, mmWave communication can combine polarization, adaptive beamforming and new spatial-processing techniques such as massive MIMO. Recent advances in the design of integrated circuits and high-directivity antennas confirm the availability of appropriate hardware solutions for achieving high gain with mmWave frequencies [12, 13].

3.3.2   MU-MIMO with Large Antenna Arrays

In dense small-cell networks, MU-MIMO technology can be used to provide back-haul connectivity to a large number of small cells simultaneously. Also, to ensure a high data rate for mmWave backhaul links, MU-MIMO with a large number of antennas is highly beneficial. Ideally, there can be an infinite number of antennas that can support an infinite number of users at the same time [3]. However, in practice, the number of serving users is limited by the finite coherence time of the channel. The increased number of antennas effectively mitigates the fast fading and uncorrelated noise; however, the impact of interference cannot be ignored. The primary types of interference in a massive MIMO system include the following:

- *Pilot contamination:* In a massive MIMO system, orthogonal pilot sequences are typically used to train the massive MIMO BSs via uplink transmissions of the users. Due to uplink–downlink reciprocity in a time-division duplex (TDD) massive MIMO system, the estimated channel information in the uplink is exploited for the downlink transmissions. Note that the fundamental limitation of the massive MIMO BSs arises from the reuse of pilot sequences in the neighbouring BSs, that is, transmissions of users in the neighbouring cells on the same pilot sequence. This pilot reuse contaminates the channel estimates of a user in a given cell. This phenomenon is referred to as coherent inter-cell interference (or pilot contamination). Nonetheless, even in the presence of pilot contamination, massive MIMO BSs can serve a large number of users simultaneously, with asymptotic SIR coverage probability given in closed form in [14].

Note that the impact of pilot contamination tends to be more severe in multi-tier networks; therefore, efficient pilot decontamination schemes will be of significant importance. Some of the pioneering works on pilot decontamination include [15] where multicast transmissions are used to eliminate the effects of pilot contamination.

- *Traditional inter-cell interference:* Apart from pilot contamination, a user rate in a massive MIMO system is also affected by the traditional inter-cell interference (i.e., non-coherent interference).
- *Intra-cell or multi-access interference:* In MU-MIMO, a BS serves multiple users simultaneously using spatial multiplexing. Thus, users form clusters within a cell where they share the same scattering environment with similar spatial channel correlations. Due to imperfect channel estimation and, in turn, imperfect precoding matrices, one communication stream can interfere with another stream within the same cluster. Nonetheless, efficient precoding/beamforming schemes can eliminate intra-cell interference in MU-MIMO systems.

With the increasing number of antennas in massive MIMO systems, the computational complexity for designing the precoders also increases. In this regard, there are several research works that aim to reduce the complexity of precoding design for massive MIMO systems. A hierarchical interference mitigation technique proposed in [16] utilizes an inner precoder for intra-cell spatial multiplexing and an outer precoder for inter-cell interference cancellation. The iterative algorithms for precoding can reduce the computational complexity significantly for massive MIMO systems. In [17], the authors propose a low-complexity precoding technique based on truncated polynomial expansion (TPE) that provides a better approximation of matrix inversion and can be optimized to maximize the weighted max–min fairness for massive MIMO systems.

3.4   MmWave Backhauling: State of the Art and Research Issues

Moving toward dense small cells in the 5G network requires a combination of backhauling techniques that may vary according to the locations, target QoS requirements and traffic load of the different SBSs. For example, indoor SBSs can obtain high-capacity backhauls from the existing wired infrastructure. By contrast, it can be more complex and expensive to set up wired backhaul connectivity for outdoor SBSs as they can possibly be installed below roof-top level, on exterior walls of buildings, street lamps or other street fixtures. In such scenarios, wireless backhauling solutions can serve the purpose. In this section, we will discuss and review the state of the art of LOS and non-LOS mmWave backhauling in terms of feasibility, topology and implementation and will then detail the fundamental challenges that exist in massive-MIMO-enabled mmWave backhaul systems.

3.4.1 LOS mmWave Backhauling

3.4.1.1 Point-to-Point (PtP) Topology

PtP topology is the traditional method of wireless backhauling in cellular networks. The mmWave frequency ranging from 60–90 GHz can be used to form PtP beams mitigating the deleterious effects of oxygen/molecular absorption and rain attenuation. Due to the limitation of mmWave propagation, the inter-site distance of picocells needs to be within 100 m for reliable PtP X2-based backhaul links [18]. The mmWave frequencies allow the formation of two or more PtP links at the same location using highly directive antenna arrays.

3.4.1.2 Point-to-Multipoint (PtMP) Topology

PtMP topology could be another suitable alternative to PtP backhauling. The PtMP wireless links are based on a hub and remote concept [19]. For example, a small cell with fibre-based backhaul connection can serve as a wireless backhaul hub and can support the backhaul transmission of six to eight small cells at a time. A PtMP in-band backhauling system was proposed in [20], where mmWave spectrum was used in both the backhaul and access links. A time-division multiplexing (TDM)-based scheduling algorithm was proposed for PtMP mmWave backhauling, where SBSs are partitioned into three sectors. In this scheduling algorithm, the backhaul links and access links are scheduled simultaneously in an adaptive manner and the hub steers beams toward the neighbouring SBSs in one sector in each time slot.

3.4.1.3 Mesh Topology

In the case of some physical obstructions, the small cells might need multiple links for reliable backhauling; therefore, the flexibility of the backhauling system can be improved further by using a multi-hop mesh network. In the multi-hop mesh topology, long-distance backhaul links are replaced by multiple short links, ensuring the reliability of the backhauling system. However, the processing and accessing delays at each hop may affect the performance of the backhauling links.

Recently, [5] considered a flexible mesh connectivity in mmWave backhauls with a strict bound on latency. Electrically steerable antenna arrays are used in this directed-mesh connection operating at mmWave frequencies. Each node is capable of self-tuning its parameters to obtain an optimal path that provides maximal throughput and minimal latency. To achieve this optimal performance, a TDM-based joint scheduling and routing algorithm is implemented where the link parameters are updated in real time. The joint scheduling of transmission over access and backhaul links using mmWave frequency maximizes the spatial reuse while managing the intra-cell and inter-cell interference efficiently [21]. In addition to the densely deployed SBSs, the density of the wireless devices can also be very high in a 5G network. For such a scenario, the centralized MAC scheduling algorithm proposed in [21] suggests enabling direct device-to-device (D2D) transmission for optimal path selection and concurrent transmission scheduling over the access link and backhaul links. In some cases, their proposed algorithm can achieve near-optimal performance in terms of throughput and latency.

The self-backhauling architecture proposed in [22] demonstrates an improved multi-hop mesh networking where a fraction of SBSs have wired backhaul and others are backhauled wirelessly. Each wired SBS provides backhaul links to multiple SBSs using mmWave frequency without imposing any interference. The authors characterized the mmWave network as noise-limited where the interference power does not cause any harm for a moderate density of SBSs. The authors analysed network performance in terms of coverage rate for different combinations of the fraction of wired backhauled SBSs and mmWave backhauled SBSs. The results presented in [22] show that increasing the fraction of wired backhauled SBSs can improve the coverage rate significantly. However, if the density of the wired backhauled SBSs is kept constant, then the rate will eventually saturate for increasing density of the wireless backhauled SBSs. In the same self-backhauling mmWave network, the authors also investigated the impact of the co-existence of an ultra-high-frequency (UHF) network with a mmWave network.

3.4.2   NLOS mmWave Backhauling

In practice, it is likely that LOS backhaul links will be blocked by buildings or other surrounding objects. This renders the use of the mmWave frequency for backhauling more difficult. Also, such links are susceptible to rain attenuation, oxygen absorption and beam misalignment (due to wind, vibration and other environmental factors). As such, it is important to have accurate directed beamforming capability with high gains and subtle beam alignment for mmWave backhaul links. In the case of non-LOS wireless backhaul links, the diffracted ray gives the propagation loss for the desired link and other reflected rays are treated as interfering links. To determine the gain for the desired link, the antenna array for each link is steered toward the point of diffraction. The mmWave frequency therefore enables a very narrow beam with high antenna gain, which, in principle, diminishes the spatial interference. However, the interference cannot be completely ignored in a scenario with ultra-dense deployment of small cells where the probability of spatial interference is high.

The NLOS PtP backhauling model considered in [23] includes the effect of rain, oxygen absorption (for 60 GHz) and antenna misalignment. For their system model, the simulation results show that the high-frequency links (60 and 73 GHz) form high-gain narrower beams with a fading and implementation margin that can compensate for additional propagation losses (due to rain and other factors) to some extent. In [18], the authors proposed a high-gain beam-alignment technique using a hierarchical beamforming codebook which is computationally efficient. Their proposed framework adaptively samples the subspace and forms an optimal beam that maximizes the received SNR. In order to validate their framework they also investigated the wind effects on beam alignment using pole movement analysis. Their analysis also shows how frequently beam alignment needs to be performed. The large antenna arrays are sensitive to beam misalignment. Therefore, more research work on mmWave backhaul links is required to investigate the trade-off between array size and achievable beamforming gain.

3.4.3   Research Challenges for Backhauling in 5G Networks

As mentioned earlier, for ultra-dense 5G networks, the mmWave frequency is envisioned as a key technology that can provide Gbps backhaul connectivity due to ample spectrum resources available in this frequency band. Additionally, due to highly directive beamforming gains, the integration of massive MIMO to the backhauling infrastructure can further enhance the reliability of the mmWave wireless backhaul links. Nonetheless, the successful roll-out of the massive MIMO and mmWave technologies for wireless backhauling is hindered by several design and implementation issues. To this end, in this section, we discuss some of the existing and anticipated research issues in the design of massive-MIMO-enabled mmWave systems.

3.4.3.1   Provision of Simultaneous Backhaul to Multiple SBSs

In ultra-dense wireless-backhauled small-cell networks, system operators need to support the backhauls of several SBSs at the same time. This requires a pool of spectrum resources that can be efficiently allocated to various small cells at the same time such that the interference among backhaul streams remains below a prescribed limit. In this context, the ample amount of spectrum in the mmWave bands and its noise-limited nature could potentially serve the purpose, especially when closely located SBSs need to be backhauled. On the other hand, massive-MIMO-based backhauling is another technique which could potentially support the backhauls of multiple SBSs within the coverage of a massive-MIMO-enabled backhaul hub. This solution is more suitable for backhaul transmissions between SBSs and the core network since it exploits PtMP transmissions in the same time and frequency resource.

For multi-user mmWave systems, multiple beams need to be formed at the same time, which necessitates efficient precoding schemes. Also, developing multi-user hybrid analogue–digital precoding for mmWave is very challenging since it requires more processing at the digital layer to manage inter-cell/intra-cell interference [24]. mmWave transceivers are thus expected to be expensive and complex in design. Recently, a digitally controlled phase shifter network (DPSN)-based hybrid precoding/combining scheme proposed in [25] was shown to be capable of reducing the cost and complexity of mmWave transceivers.

3.4.3.2   Acquisition of Channel State Information (CSI)

With multi-user massive MIMO technology and efficient beamforming, it becomes possible to serve the backhauls of a large number of SBSs with large degrees of freedom. However, CSI estimation is an underlying requirement which strongly impacts the performance of beamforming in massive MIMO systems. Traditionally, in massive MIMO systems, the channels between transmitters and receivers are estimated from orthogonal pilot sequences. These sequences are, however, limited in number due to the finite coherence time of the channel. Therefore, it becomes crucial to reuse the same pilot sequences in a multi-tier network. This reuse of pilot sequences in different cells leads to pilot contamination that limits the rate gains of a massive MIMO system. To overcome this issue, coordinated multi-point transmissions (CoMP) can be employed. Also, a set of antenna elements in a massive MIMO system can be leveraged to mitigate the impact of pilot contamination.

3.4.3.3   Adaptive Backhaul/Access Spectrum Selection

Traditionally, the network operators optimize the subchannels of a typical RF spectrum (sub-6 GHz) that are allocated to a given user at its access links given the wired backhaul at BSs. However, the possible use of a combination of frequency bands in 5G access/backhaul networks such as microwave, mmWave and sub-6 GHz renders this task particularly challenging. The reason is that the incurred interference and offered network capacity at different frequency bands can be different for various indoor/outdoor environments. For instance, compared to the traditional sub-6 GHz band, the mmWave frequencies have high penetration/attenuation losses that can vary significantly for indoor and outdoor propagation environments. A careful system-level analysis is therefore required to adaptively select an appropriate combination of frequencies for access/backhauls taking into account the crucial factors such as interference conditions at a given frequency, locations of the SBSs, their surrounding environment, transmit/receive BS antenna characteristics and the beamforming gains.

Moreover, the possibility of reusing the same frequency at both the backhaul and access links (i.e., in-band backhauling) with minimal interference is also crucial and can vary for different spectrum bands. For example, the high directivity and noise-limited nature of mmWave spectrum can strongly support in-band backhauling compared to traditional sub-6 GHz frequency spectrum. Consequently, choosing a feasible spectrum band that can enhance data rates via in-band backhauling is also crucial.

3.4.3.4   Backhaul Spectrum-aware User Association

As mentioned previously, different frequency bands may result in significantly varying data rates even in similar system settings. Therefore, *optimal spectrum selection* to serve a user under given circumstances is crucial and requires the operation of SBSs in a variety of frequency bands [26]. Nonetheless, deploying such SBSs may not be a trivial task due to hardware modifications as well as deployment costs. Therefore, from the perspective of network operators, the significance of determining an efficient, low-complexity, traffic-offloading criterion is evident. Also, from the perspective of users, it is important to choose a user-association criterion that considers propagation losses for the frequency used and antenna-specific parameters of the SBSs.

3.4.3.5   Backhaul/Access Link Scheduling

In practice, the massive-MIMO-enabled backhaul system can serve a limited number of backhaul streams at a given time and frequency resource. Consequently, the significance of backhaul scheduling becomes evident in dense cellular networks. In addition to traditional time-division-based scheduling, more sophisticated scheduling techniques can be implemented for mmWave massive MIMO networks. For example, in [25], the proposed beam-division-multiplexing (BDM)-based scheduling is shown to improve performance gain for in-band mmWave backhauling. Therefore, the combination of TDM and BDM can make the user-scheduling procedure much more flexible. Moreover, unlike traditional user scheduling, the backhaul scheduling requires consideration of the traffic load of a serving SBS and average achievable data rate at the access links in addition to the backhaul channel conditions.

3.4.3.6   Number of RF Chains

Typically, MIMO systems are equipped with few antennas and, in turn, the number of employed RF chains, digital-to-analogue converters (DACs) and analogue-to-digital converters (ADCs) can be comparable to the number of antennas. However, in a massive MIMO system, deploying RF chains in a comparable quantity is not practically feasible. Due to this limitation, the orthogonal channel estimation approach may not be used even without pilot contamination. Also, the energy consumption of a MIMO transceiver increases as a function of the number of active RF chains due to the high energy consumption of ADCs. Thus, the channel estimation and beamforming algorithms should be designed taking into account the constraints on the number of RF chains.

3.4.3.7   Other Implementation Issues

The directed multi-gigabit (DMG) PHY specification in the IEEE standard 802.11ad suggests OFDM modulation for high-data-rate applications. As mmWave propagation characteristics are quite different from those for the microwave, the OFDM parameters for 3GPP LTE need to be modified. For mmWave communication, OFDM PHY has different frame structures and can be implemented using QPSK, 16-QAM and 64-QAM. The bandwidth of OFDM subcarriers and guard intervals needs to be larger. The basic TDD-based mmWave frame has 10 subframes and each subframe contains 14 OFDM symbols [27]. When mmWave is used for wireless backhauling, subframes can be configured to support multi-hop transmission utilizing spatial multiplexing [27]. The configuration of mmWave subframing needs further investigation.

3.5   Case Study: Massive-MIMO-based mmWave Backhauling System

As depicted in Figure 3.1, we consider wireless backhaul hubs installed within macro base stations (MBSs) to provide backhaul connectivity to SBSs or access points (APs) over mmWave links utilizing massive MIMO technology. The APs use different mmWave frequencies in the access and backhaul links. We assume that the APs and user equipment (UEs) are equipped with directional antennas with sectorized gain pattern. Each hub supports N_b APs simultaneously with M_h antennas. Within the same time block, each AP also schedules N_a UEs. The transmission power of a hub and AP are P_h and P_a, respectively. During the training phase, each AP sends a preassigned orthogonal pilot sequence to the hub, which is estimated perfectly by the hub and the pilot sequence is not used by any other neighbouring hub (i.e., no pilot contamination is assumed). The channel estimation is facilitated with time-division duplexing at the hubs so that the channel reciprocity is guaranteed.

Since the walls of buildings are impenetrable to mmWave signals, indoor APs can neither serve outdoor users nor interfere with outdoor AP transmission. In the outdoor environment, the APs are deployed on the exterior of buildings or street fixtures where users are more likely to be NLOS to the hubs and users. We focus on outdoor user performance and assume that the hub–AP and AP–user channels fade independently across time slots.

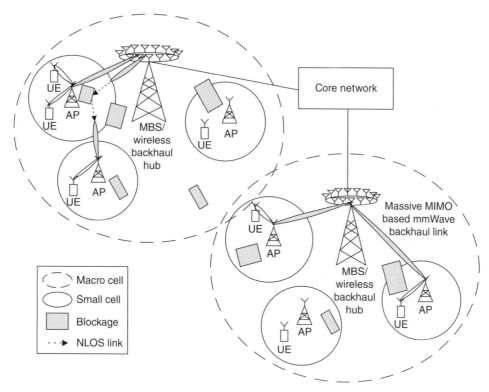

Figure 3.1 Small-cell network architecture incorporating massive-MIMO-based mmWave backhauling system

3.5.1 System Model

We consider that backhaul hubs and APs are distributed according to homogeneous Poisson point processes (PPPs) in a cellular region $\mathbb{R}^2$. The PPPs formed by the hubs and APs are $\Phi_\mathrm{H} = \{H_i\}$ with density λ_H and $\Phi_\mathrm{A} = \{A_j^{(i)}\}$ with density λ_A, respectively. The locations of UEs are approximated by a PPP $\Phi_\mathrm{U} = \{U_k^{(j)}\}$ with density λ_U. We denote by $A_j^{(i)}$ the jth AP backhauled by the ith hub and $U_k^{(j)}$ as the kth UE served by the jth AP. The buildings and other outdoor blockages are modelled using the Boolean scheme of rectangles [28]. In this model, the blockages are considered as a process of random rectangles where the centres of the rectangles form a PPP Φ_b with density λ_b. The length, width and orientation of the rectangles are determined as independent and identically distributed (i.i.d.) random variables. We also assume that the distribution of blockages is stationary and does not vary with any translation or rotation. Over the geographic area, the probability that a network node will be in the LOS region is determined based on the size and density of the blockages [29]. Considering downlink transmission, we denote the distance between a transmitting node $\{X_\mathrm{t}\}$ and

receiving node $\{Y_r\}$ as R_r, where $X_t \in \{\Phi_H, \Phi_A\}$ and $Y_r \in \{\Phi_A, \Phi_U\}$. The LOS probability for the link is given by:

$$p(R_l) = e^{-R_l/\rho} \tag{3.1}$$

where ρ is the average LOS distance.

3.5.1.1 Directivity Gain

We model directional beamforming gains as marks of the PPPs and approximate the
actual beamforming pattern using the sectored gain pattern presented in [22, 28, 30] for
ad hoc and mmWave networks. The antenna pattern is represented as $D_{G,g,\theta}(\phi)$, where G
is the main lobe directivity gain, g is the side-lobe directivity gain and θ is the beam width
of the main lobe with the antenna boresight direction ϕ at the node. For a given angle of
departure ϕ_t^l at the transmitter and angle of arrival ϕ_r^l at the receiver, the total directivity
gain for the link R_l is determined as $D_{t,r}^{(l)} = D_{G_t, g_t, \theta_t}(\phi_t^l) D_{G_r, g_r, \theta_r}(\phi_r^l)$. For the desired link
(l_0) between the target nodes, the directivity gain is $D_{t,r}^{(l_0)} = G_t G_r$. The number of beams
from interfering nodes is considered to be a discrete random variable.

Considering the probability distribution for four possible scenarios, the directivity
gain of the interfering links is $D_{t,r}^{(l_i)} = a_n$ with probability $b_n (n \in \{1, 2, 3, 4\})$, where we
use a_n and b_n constants as proposed in [29]. The probability distribution for the inter-
fering links is shown in Table 3.1. Here, $c_t = \dfrac{\theta_t}{2\pi}$ and $c_r = \dfrac{\theta_r}{2\pi}$. For example, $U_k^{(j)}$ is
associated with $A_j^{(i)}$ and the random distribution of interference link from $A_{j'}^{(i)}$ to $U_k^{(j)}$
in the downlink represents that the main lobe of $U_k^{(j)}$ is overlapped by a side lobe of
$A_{j'}^{(i)}$. The directivity gain of the interfering link will then be $D_{j',k}^{(l_i)} = a_2 = g_t G_k$ with
probability $(1 - c_{j'})c_k$.

3.5.1.2 Path-loss Model

MmWave links may experience either LOS or NLOS propagation. Therefore, the
path loss experienced by the link R_l can be determined as:

$$L(R_l) = \mathbb{I}(p(R_l)) K_L R_l^{-\alpha_L} + (1 - \mathbb{I}(p(R_l))) K_N R_l^{-\alpha_N} \tag{3.2}$$

Table 3.1 Probability distribution function of $D_{t,r}^{(l_i)}$

i	1	2	3	4
a_i	$G_t G_r$	$g_t G_r$	$G_t g_r$	$g_t g_r$
b_i	$c_t c_r$	$(1 - c_t)c_r$	$c_t(1 - c_r)$	$(1 - c_t)(1 - c_r)$

where $\mathbb{I}(p(R_l))$ is the Bernoulli RV for the LOS probability $p(R_l)$ and α_L and α_N are the LOS and NLOS path-loss exponents. The log-normal shadowing and path loss at a reference distance are approximated as K_L and K_N for LOS and NLOS, respectively. We assume small-scale fading h_l as a normalized gamma random variable.

3.5.1.3 User Association

We consider that each UE can associate with one AP which is backhauled by at most one hub. The association between AP and UE is indicated by:

$$\gamma_{kj} = \begin{cases} 1, & \text{if } U_k^{(j)} \text{associated with } A_j^{(i)} \\ 0, & \text{otherwise.} \end{cases}$$

Resources at the APs are uniformly allocated to a maximum of Q_a UEs according to the effective load. The percentage of resource consumption for a UE is denoted as $\beta_{kj} = \dfrac{1}{\sum_{U_k^{(j)} \in \Phi_U} \gamma_{kj}}$. The hub–AP association indicator is considered as:

$$\delta_{ji} = \begin{cases} 1, & \text{if } A_j^{(i)} \text{ associated with } H_i \\ 0, & \text{otherwise.} \end{cases}$$

Resources at the hubs are uniformly allocated among a maximum of Q_h APs. The percentage of resource consumption for an AP is denoted as $\eta_{ji} = \dfrac{1}{\sum_{A_j^{(i)} \in \Phi_A} \delta_{ji}}$.

3.5.1.4 Interference Model

As we focus on a massive MIMO network with a large number of antennas implemented in hubs, where $N_b \ll M_h$, the large-scale antenna arrays diminish the effects of uncorrelated noise and the system is considered to be interference limited [3]. We assume that efficient beamforming and a pilot decontamination scheme eliminate the effects of intra-cell interference and pilot contamination. Thus, an AP will receive interference only from other hubs that are transmitting in the downlink. The interference received by $A_j^{(i)}$ (the jth AP which is backhauled by the ith hub) is expressed as follows:

$$I_b = \sum_{l_l > 0:\, H_{l'} \in \Phi_H \setminus \{H_i\}} P_h \left| h_{l_l} \right|^2 D_{i',j}^{(l_l)} L(R_{l_l}) \sum_{A_{j'}^{(i)} \in \Phi_A} \delta_{j'i'} \eta_{j'i'}. \tag{3.3}$$

The access link interference between the user element $U_k^{(j)}$ and access point $A_j^{(i)}$ can be expressed as follows:

$$I_a = \sum_{l_l > 0: A_{j'}^{(i)} \in \Phi_A \setminus \{A_j^{(i)}\}} P_a \left| h_{l_l} \right|^2 D_{j',k}^{(l_l)} L\left(R_{l_l}\right) \sum_{U_{k'}^{(j')} \in \Phi_U} \gamma_{k'j'} \beta_{k'j'}. \tag{3.4}$$

3.5.1.5   SINR and Rate Calculation

In the massive MIMO regime, the SINR at the backhaul link for AP $A_j^{(i)}$ can be approximated as follows [31]:

$$\mathrm{SINR}_b = \left(\frac{1 + M_h - N_b}{N_b}\right)\left(\frac{P_h \left| h_{l_0} \right|^2 D_{i,j}^{(l_0)} L\left(R_{l_0}\right)}{1 + I_b}\right). \tag{3.5a}$$

The SINR received at $U_k^{(j)}$ in the access link is given by:

$$\mathrm{SINR}_a = \frac{P_a \left| h_{l_0} \right|^2 D_{j,k}^{(l_0)} L\left(R_{l_0}\right)}{\sigma_{N^2} + I_a} \tag{3.5b}$$

where σ_{N^2} is the thermal noise power. The user rate in the downlink is therefore calculated as:

$$R_{kj} = B\log_2\left(1 + \min\{\mathrm{SINR}_a, \mathrm{SINR}_b\}\right) \tag{3.6}$$

where B is the bandwidth of the channel assigned to the user.

3.5.2   Maximizing User Rate

In the downlink transmission, the hubs forward the traffic to the APs over the backhaul links and then the APs forward the traffic to the desired users. The hubs can support a fixed number of APs simultaneously and APs also have the constraint to support a limited number of users at a time. Therefore, the downlink rate is greatly affected by the association at the backhaul and access links. In this context, we formulate a user-association problem for the downlink transmission to maximize the overall user rate as follows:

$$\max_{\delta_{ji}, \gamma_{kj}} \left(\sum_{A_j^{(i)} \in \Phi_A} \sum_{U_k^{(j)} \in \Phi_U} \gamma_{kj} \beta_{kj} R_{kj}\right), \tag{3.7a}$$

$$\text{s.t.} \sum_{H_i \in \Phi_H} \delta_{ji} = 1, \sum_{A_j^{(i)} \in \Phi_A} \gamma_{kj} = 1, \quad \forall A_j^{(i)} \in \Phi_A, U_k^{(j)} \in \Phi_U, \tag{3.7b}$$

$$\sum_{A_j^{(i)} \in \Phi_A} \delta_{ji} = Q_h, \sum_{U_k^{(j)} \in \Phi_U} \gamma_{kj} = Q_a, \quad \forall H_i \in \Phi_H, A_j^{(i)} \in \Phi_A. \tag{3.7c}$$

Since solving this optimization problem can be computationally complex in real time, we focus on devising less complex distributed user-association solutions. The theory of matching can be used to design efficient distributed solutions for user associations [32, 33].

3.5.3 Matching Theory for User Association

We address the combinatorial problem of user association in the downlink by transforming it to a two-sided matching game. For two disjoint sets of players in the matching game, a matching is performed between the two sets of players according to each player's preference metric. To obtain a stable matching, each player is assigned a fixed quota which is referred to as the maximum number of players that each player can be matched to. To solve the user-association problem in our system model, the association in the backhaul and access links is performed using the matching framework presented in [33–35].

3.5.3.1 Matching Game for Hub–AP Association

At first we consider a many-to-one matching game for the association between the hubs and APs. We denote the set of hubs by $\Phi_H = \{1, 2, ..., H\}$ and the set of APs by $\Phi_A = \{1, 2, ..., A\}$. Each AP can be matched with at most one hub and each hub can have matching with one or more APs depending on its quota Q_h. The preference relations for hubs and APs depend on the channel state of each backhaul link. Each AP aims to maximize its utility function given by its achievable rate. Considering a random association, each AP calculates the rate that it can receive from each hub and makes its preference list $\succ_{A_b} = \{\succ_a\}_{a \in \Phi_A}$ over the hubs. Similarly, each hub constructs its preference list $\succ_H = \{\succ_h\}_{h \in \Phi_H}$ using the rates calculated from the initial random association. Each hub gives a ranking score to each AP based on its preference list.

After constructing the preference lists, matching among the disjoint sets of hubs and APs is performed iteratively until a stable match is found. A matching μ_b that represents the hub–AP association can be expressed as a function from the set $\Phi_H \cup \Phi_A$ into the set $\Phi_H \cup \Phi_A$ such that [33, 35]:

- $|\mu_b(a)| = 1, \forall a \in \Phi_A$ and $\mu_b(a) = a$ if $a \notin \Phi_A$
- $|\mu_b(h)| \leq Q_h, \forall h \in \Phi_H$
- $h \in \mu_b(a)$ if and only if $\mu_b(h) = a$.

Here, the matching function $\mu_b(a)$ denotes the matched APs and $\mu_b(h)$ represents the matched hubs. As described in Algorithm 1 below, the APs first apply to their most preferred hubs according to the preference list. If the quota of the hub Q_h is not overloaded, it will schedule the AP. If the quota is exceeded, the hub checks the ranking for the applicant AP. If the ranking of the AP is better than the rank of other scheduled APs, then it discards the worst AP with maximum rank and associates with the applicant AP. This will continue until all APs are associated and their preference lists become empty. Thus, the algorithm will end with a stable matching μ_b^*.

Algorithm 1: Matching algorithm for hub–AP association

```
1.  input: Φ_H, Φ_A and Q_h
2.  initialization: calculate the preference lists ≻_H and ≻_A_b,
       respectively
3.  each hub h∈Φ_H gives a ranking score to APs based on ≻_H
4.  while (at least one AP is free AND its preference list ≻_A_b
       is not empty) do
5.       each unassociated AP applies to its most preferred hub
            in ≻_A_b
6.       if the Q_h is not overloaded then
7.          associate with the applicant AP
8.       else
9.          compare the ranking of the applicant with the
            currently associated APs
10.         if rank (applicant AP) < rank (associated APs) then
11.            discard the worst AP from the hub's current associations
12.            associate with the applicant AP
13.            discard the hub from the preference list of the
               discarded AP
14.            set the discarded AP as free
15.         end if
16.      end if
17. end while
18. output: μ_b^*
```

3.5.3.2   Matching Game for AP–UE Association

Once we have the stable matching μ_b^* for the backhaul connectivity, we perform another many-to-one matching game to find association among APs and UEs at the access end. We consider the set of APs as $\Phi_A = \{1,2,....,A\}$ and the set of UEs as $\Phi_U = \{1,2,....,U\}$. This matching game aims to match each UE with one AP and each hub with one or more UEs depending on its quota Q_a. Using the stable matching μ_b^* at the backhaul end and a random association at the access end, each UE calculates the rate that it can receive from each AP using Equation (3.6) and makes its preference list $\succ_U = \{\succ_u\}_{u \in \Phi_U}$ over the APs. APs also construct their preference list

$\succ_A = \{\succ_a\}_{a \in \Phi_A}$ depending on the total rate they can provide to the associated users. Each AP also gives a ranking score to UEs based on its preference. To find a stable matching μ_a^* for the AP–UE association, matching is performed iteratively as described in Algorithm 2. In this case, a matching μ_a representing the AP–UE association is expressed as a function from the set $\Phi_A \cup \Phi_U$ into the set $\Phi_A \cup \Phi_U$ such that:

- $|\mu_a(u)| = 1$, $\forall u \in \Phi_U$ and $\mu_a(u) = u$ if $u \notin \Phi_U$
- $|\mu_b(a)| \leq Q_a$, $\forall a \in \Phi_A$
- $a \in \mu_b(u)$ if and only if $\mu_b(a) = u$.

Here, the matching function $\mu_b(u)$ denotes the matched UEs and $\mu_b(a)$ represents the matched APs. The two-sided matching algorithms to find the hub–AP and AP–UE association are guaranteed to converge as the applicant (AP or UE) never applies to its preferred node (hub or AP) twice and thus the algorithm will have a finite number of iterations [32]. The algorithms return the stable matching μ_b^* and μ_a^* whenever any unassociated AP or UE has no hub or UE left in its preference list, respectively.

Algorithm 2:   Matching algorithm for AP–UE association

```
1. input: Φ_A, Φ_U and Q_a
2. initialization: calculate the preference lists ≻_A and ≻_U,
   respectively
3. each AP a ∈ Φ_A gives a ranking score to UEs based on ≻_A
4. while (at least one UE is free AND its preference list ≻_U
   is not empty) do
5.    each free UE U_k applies to its most preferred AP in ≻_U
6.    if the Q_a is not overloaded then
7.       associate with the applicant UE
8.    else
9.       compare the ranking of the applicant with the
         currently associated UEs
10.      if rank(applicant UE) < rank(associated UEs) then
11.         discard the worst UE from the AP's current
            associations
12.         associate with the applicant UE
13.         discard the AP from the preference list of the
            discarded UE
14.         set the discarded UE as free
15.      end if
16.   end if
17. end while
18. output: μ_a*
```

3.5.4   Numerical Results

In this section, we present numerical results for the performance of a typical user in the downlink with traditional distance-based association and distributed stable matching algorithms. The massive-MIMO-based mmWave backhaul hub is assumed to be equipped with 256 antennas and operates at 60 GHz with channel bandwidth of 2 GHz. The AP operates at 73 GHz and the channel bandwidth is 2 GHz. The transmission power for the hub and AP are 46 dBm and 30 dBm, respectively. To approximate the directivity gains for the hubs, AP and UE, we consider a 10 degree beam width for the hub and AP with a main-lobe gain of 18 dB and a side-lobe gain of −4 dB for both the mmWave hubs and APs. The directive beamforming for users is approximated by 10 dB beams with 90 degree beam width. We assume the average LOS distance $\rho = 141.4$ m. The path-loss exponents for LOS and NLOS links (both access and backhaul) are taken as 2 and 3.5, respectively.

First, in Figure 3.2, we compare the SINR coverage probability for a massive-MIMO-based mmWave backhaul system for the backhaul link and access link with both microwave and mmWave frequencies. Here, we consider $\lambda_H = 5$, $\lambda_A = 100$, $\lambda_U = 300$ per sq. km with $Q_h = 10$ and $Q_a = 4$. For fair comparison, we include similar blockage for microwave transmission. As shown in Figure 3.2, mmWave frequencies offer a better SINR coverage than microwave transmission in the access links.

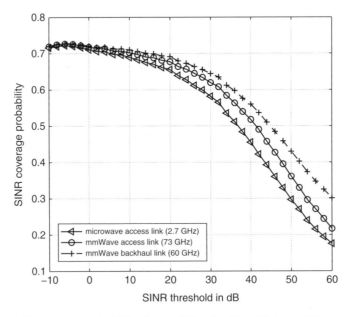

Figure 3.2   SINR coverage probability for mmWave backhaul link, mmWave and microwave access link

Since mmWave transmission uses a wider channel bandwidth than microwave transmission, the user rate will be higher for mmWave communication. The comparison of average user rate between mmWave and microwave transmission shown in Figure 3.3 indicates that microwave transmission is unable to provide a Gbps data rate to the user end as expected from 5G small-cell networks. The mmWave frequency can provide a multi-Gbps data rate on average even for densely deployed SBSs. We also observe that the user rate depends on the backhaul quota (Q_h) or maximum limit of the hubs. In Figure 3.3, it is shown that a higher average user rate can be achieved by increasing the backhaul quota which enables hubs to provide more backhaul links to the APs.

To analyse the impact of association schemes on user rates, we compare the downlink rates of all users for conventional nearest AP association and stable matching association. Figure 3.4 shows the total network rate and average user rates for different backhaul quotas (Q_h) and association schemes. Here, we consider $\lambda_H = 5$ and $\lambda_A = 100$ with AP quota $Q_a = 3$ for simulation. We observe that when the competition for AP–UE association is less for lower user density, the nearest AP association performs better than the stable-matching-based association. As user density increases, the competition for association with the desired AP also increases. In such scenarios, the overall user rate increases for the stable matching association when compared to the distance-dependent association due to several factors.

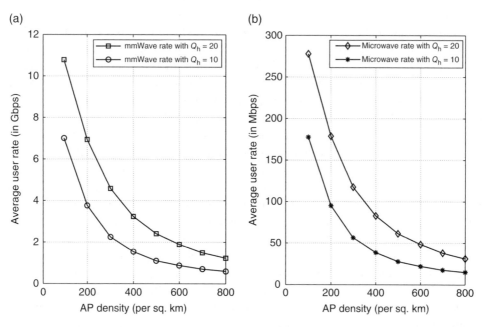

Figure 3.3 Average user rate for (a) mmWave and (b) microwave access link with different backhaul quotas (Q_h)

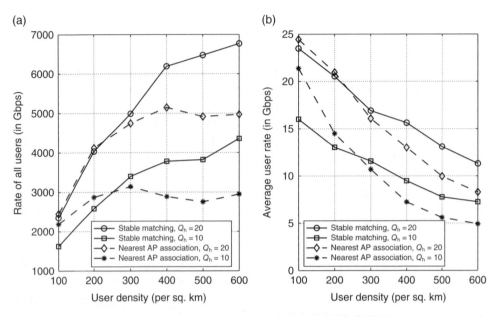

Figure 3.4   (a) Network rate and (b) average user rate for different user-association schemes

In the case of distance-dependent backhaul link association, densely deployed APs compete to get associated with the nearest backhaul hub. The hub associates with the APs according to the arrival of the association requests. Once the quota is exceeded, the hub does not have the flexibility to alter the association even if it knows that a better rate can be achieved. An AP cannot get backhaul connectivity if the nearest hub is overloaded. Similarly, for access link association, the UEs send their association requests to their nearest APs and get associated if the AP quota is not exceeded. When the competition for user association increases, more users are rejected. Also, the distance-dependent association may result in poor SINR at the receivers in some scenarios. For example, if the target AP's nearest backhaul hub is NLOS or quite far away, then downlink transmission to the AP may experience poor SINR.

On the other hand, stable matching is performed utilizing the channel state information. The preference lists for hubs, APs and UEs are calculated based on the rate information. During backhaul link association, the stable matching algorithm allows the overloaded hub to recheck the ranking or the rates it can deliver to APs it is currently associated with. The hub can update its association if it can provide a better rate to the requesting AP. The stable matching algorithm also provides an opportunity for an AP to apply to the next preferred hub if it is rejected by its most preferred hub. The next preferred hub may be able to support this requesting AP with a better rate. Consequently, this will allow the AP to serve more UEs in the network. Similarly, the stable matching algorithm for the AP–UE association, which exploits the channel state information, will also enable UEs to achieve better user rates at the access links when compared to those with the nearest AP association.

3.6 Conclusion

We have discussed wireless backhaul solutions for future 5G small-cell networks. In order to provide backhaul connectivity to a large number of SBSs in such a network in a cost-effective way, mmWave spectrum is beneficial due to the ample amount of unlicensed spectrum and high directivity. The performance of mmWave propagation can be enhanced by using massive MIMO through installing a large number of antennas in the backhaul system. We have described a tractable system model for a massive-MIMO-enabled mmWave backhauling network. To maximize the overall user rate, we have formulated a matching game which provides stable matching association at the backhaul and access ends. Numerical results have shown that the massive-MIMO-enabled mmWave backhaul solution provides a reliable SINR coverage probability and the use of mmWave in the access link increases the average user rate compared to traditional microwave. Analysis of the stable matching user association has shown that channel-state-aware matching can provide a better user rate compared to distance-dependent user association in a dense small-cell network.

The insights from the results motivate the design of a backhaul-aware user association scheme so that the overall network is maximized in a dense small-cell network. To improve the network performance further, massive MIMO can be incorporated at the APs so that more users can be served simultaneously with higher antenna gain.

Acknowledgement

This work was supported by the Natural Sciences and Engineering Research Council of Canada (NSERC).

References

[1] Andrews, J. G., Buzzi, B., Choi, W., Hanly, S. V., Lozano, A., Soong, A. C. K. and Zhang, J. C. (2014) What will 5G be?. *IEEE Journal on Selected Areas in Communication*, **32**(6), 1065–1082.

[2] Rangan, S., Rappaport, T. and Erkip, E. (2014) Millimeter-wave cellular wireless networks: Potentials and challenges. *Proceedings of the IEEE*, **102**(3), 366–385.

[3] Marzetta, T. L. (2010) Noncooperative cellular wireless with unlimited numbers of base station antennas. *IEEE Transactions on Wireless Communications*, **9**(11), 3590–3600.

[4] Siddique, U., Tabassum, H., Hossain, E. and Kim, D. I. (2015) Wireless backhauling of 5G small cells: Challenges and solution approaches. *IEEE Wireless Communications, Special Issue on 'Smart Backhauling and Fronthauling for 5G Networks'*, **22**(5), 22–31.

[5] Interdigital (2013) 'Small Cell Millimeter Wave Mesh Backhaul.' White paper. Interdigital, Wilmington, DE, USA. Available at: http://goo.gl/Dl2Z6V.

[6] Li, Y., Pappas, N., Angelakis, V., Pioro, M. and Yuan, D. (2015) Optimization of free space optical wireless network for cellular backhauling. *IEEE Journal on Selected Areas in Communication*, **33**(9), 1841–1854.

[7] Demers, F., Yanikomeroglu, H. and St-Hilaire, M. (2011) A survey of opportunities for free space optics in next generation cellular networks. In *Proceedings of the Communication Networks and Services Research Conference*, pp. 210–216, May.

[8] Zhao, Q. and Li, J. (2006) Rain attenuation in millimeter wave ranges. In *Proceedings of the IEEE International Symposium on Antennas, Propagation and EM Theory*, pp. 1–4, October.

[9] Rappaport, T., Sun, S., Mayzus, R., Zhao, H., Azar, Y., Wang, K., Wong, G., Schulz, J., Samimi, M. and Gutierrez, F. (2013) Millimeter wave mobile communications for 5G cellular: It will work! *IEEE Access*, **1**, 335–349.

[10] Daniels, R. C. and Heath, R. W. (2007) 60 GHz wireless communications: Emerging requirements and design recommendations. *IEEE Vehicular Technology Magazine*, **2**(3), 41–50.

[11] Roh, W., Seol, J., Park, J., Lee, B., Lee, J., Kim, Y., Cho, J., Cheun, K. and Aryanfar, F. (2014) Millimeter-wave beamforming as an enabling technology for 5G cellular communications: Theoretical feasibility and prototype results. *IEEE Communications Magazine*, **52**(2), 106–113.

[12] Rappaport, T. S., Murdock, J. N. and Gutierrez, F. (2011) State of the art in 60-GHz integrated circuits and systems for wireless communications. *Proceedings of the IEEE*, **99**(8), 1390–1436.

[13] Dussopt, L., Bouayadi, O. E., Luna, J. A. Z., Dehos, C. and Lamy, Y. (2015) Millimeter-wave antennas for radio access and backhaul in 5G heterogeneous mobile networks. *European Conference on Antennas and Propagation*, pp. 1–4, April.

[14] Bai, T. and Heath, R. W. (2014) Asymptotic coverage probability and rate in massive MIMO networks. In *Proceedings of the IEEE GlobalSIP*, pp. 602–606, December.

[15] Xiang, Z., Tao, M. and Wang, X. (2014) Massive MIMO multicasting in noncooperative cellular networks. *IEEE Journal on Selected Areas in Communication*, **32**(6), 1180–1193.

[16] Liu, A. and Lau, V. (2014) Hierarchical interference mitigation for massive MIMO cellular networks. In *IEEE Transactions on Signal Processing*, **62**(18), 4786–4797.

[17] Kammoun, A., Muller, A., Bjornson, E. and Debbah, M. (2014) Low-complexity linear precoding for multi-cell massive MIMO systems. In *Proceedings of EUSIPCO*, pp. 2150–2154, September.

[18] Hur, S., Kim, T., Love, D. J., Krogmeier, J. V., Thomas, T. A. and Ghosh, A. (2013) Millimeter wave beamforming for wireless backhaul and access in small cell networks. *IEEE Transactions on Communication*, **61**(10), 4391–4403.

[19] Wireless 20/20 (2012) 'Rethinking Small Cell Backhaul: A Business Case Analysis of Cost-Effective Small Cell Backhaul Network Solutions.' White paper, July. Available at: http://www.wireless2020.com/docs/RethinkingSmallCellBackhaulWP.pdf

[20] Taori, R. and Sridharan, A. (2015) Point-to-multipoint in-band mmWave backhaul for 5G networks. *IEEE Communications Magazine*, **53**(1), 195–201.

[21] Niu, Y., Gao, C., Li, Y., Su, L., Jin, D. and Vasilakos, A. V. (2015) Exploiting device-to-device transmissions in joint scheduling of access and backhaul for small cells in 60 GHz band. *IEEE Journal on Selected Areas in Communication*, **33**(10), 2052–2069.

[22] Singh, S., Kulkarni, M. N., Ghosh, A. and Andrews, J. G. (2015) Tractable model for rate in self-backhauled millimeter wave cellular networks. *IEEE Journal on Selected Areas in Communication*, **31**(10), 2196–2211.

[23] Coldrey, M., Koorapaty, H., Berg, J., Ghebretensa, Z., Hansryd, J., Derneryd, A. and Falahati, S. (2012) Small-cell wireless backhauling: A non-line-of-sight approach for point-to-point microwave links. In *Proceedings of the IEEE Vehicular Technology Conference*, pp. 1–5, September.

[24] Alkhateeb, A., Mo, J., Gonzalez-Prelcic, N. and Heath, R. W. (2014) MIMO precoding and combining solutions for millimeter-wave systems. *IEEE Communications Magazine*, **52**(12), 122–131.

[25] Gao, Z., Dai, L., Mi, D., Wang, Z., Imran, M. A. and Shakir, M. Z. (2015) MmWave massive-MIMO-based wireless backhaul for the 5G ultra-dense network. *IEEE Wireless Communications*, **22**(5), 13–21.

[26] Wang, N., Hossain, E. and Bhargava, V. K. (2015) Backhauling 5G small cells: A radio resource management perspective. *IEEE Wireless Communications*, **22**(5), 41–49.

[27] Zheng, K., Zhao, L., Mei, J., Dohler, M., Xiang, W. and Peng, Y. (2015) 10Gbps-HetSNets with millimeter-wave communications: Access & networking challenges and protocols. *IEEE Communications Magazine*, **53**(1), 222–231.

[28] Bai, T., Vaze, R. and Heath, R. W. (2014) Analysis of blockage effects on urban cellular networks. *IEEE Transactions on Wireless Communication*, **13**(9), 5070–5083.

[29] Bai, T. and Heath, R. W. (2015) Coverage and rate analysis for millimeter-wave cellular networks. *IEEE Transactions on Wireless Communication*, **14**(2), 1100–1114.

[30] Hunter, A., Andrews, J. and Weber, S. (2008) Transmission capacity of ad hoc networks with spatial diversity. *IEEE Transactions on Wireless Communication*, **7**(12), 5058–5071.

[31] Bethanabhotla, D., Bursalioglu, O., Papadopoulos, H. C. and Caire, G. (2014) User association and load balancing for cellular massive MIMO. In *Proceedings of Information Theory and its Applications (ITA)*, pp. 1–10, February.

[32] Roth, A. E. and Sotomayor, M. A. O. (1992) Two-sided matching: A study in game-theoretic modeling and analysis, Cambridge University Press, pp. 485–541.

[33] Jorswieck, E. (2011) Stable matchings for resource allocation in wireless networks. In *Proceedings of the 17th International Conference on Digital Signal Processing (DSP)*, pp. 1–8, July.

[34] Semiari, O., Saad, W., Dawy, Z. and Bennis, M. (2015) Matching theory for backhaul management in small cell networks with mmWave capabilities. In *Proceedings of the IEEE ICC*, pp. 3460–3465, June.

[35] Sekander, S., Tabassum, H. and Hossain, E. (2015) A matching game for decoupled uplink–downlink user association in full-duplex small cell networks. In *IEEE Globecom'15*, December.

4

Fronthaul for a Flexible Centralization in Cloud Radio Access Networks

Jens Bartelt,[1] Dirk Wübben,[2] Peter Rost,[3] Johannes Lessmann[4] and Gerhard Fettweis[1]

[1] *Technische Universität Dresden, Dresden, Germany*
[2] *University of Bremen, Bremen, Germany*
[3] *Nokia Networks, Munich, Germany*
[4] *NEC Laboratories Europe, Heidelberg, Germany*

4.1 Introduction

The architecture of fourth generation (4G) mobile networks, for example 3GPP long-term evolution (LTE), is organized in a decentralized way, such that the complete baseband processing including the physical (PHY) layer, medium access (MAC) layer and parts of the network layer processing are performed at the base stations (BSs). Internet Protocol (IP)-layer user data are then forwarded between the BS and the network core, which requires a relatively modest transport network, usually known as the backhaul (BH) network.

An alternative to this decentralized concept is to centralize radio access network (RAN) functionalities. First introduced in [1], the so-called centralized-RAN or cloud-RAN (C-RAN) architecture proposes to reduce the functionality of BSs to so-called remote radio heads (RRHs), which only perform analogue processing and forward digital samples between the RRH and centralized baseband units (BBUs). Such a centralized architecture is already utilized in some 4G networks and is actively

Backhauling/Fronthauling for Future Wireless Systems, First Edition.
Edited by Kazi Mohammed Saidul Huq and Jonathan Rodriguez.

being considered for future mobile networks as it offers several advantages. First among these advantages is a reduction in operational and capital expenditure. By reducing the size, the sites can be smaller and the energy consumption can be reduced, particularly because no active cooling is required. As all BBUs are located in a centralized location, maintenance also becomes much easier.

Next, the centralization of BBUs makes cooperative processing techniques much easier to implement. Techniques such as coordinated multi-point (CoMP) [2, 3] or multi-point turbo detection (MPTD) [4] require an extensive exchange of signals between BSs and suffer heavily from delays on BH links. Hence, they are much easier to implement if all signals are forwarded and processed in a centralized BBU.

Finally, current advances in processor technology and virtualization have enabled the implementation of baseband processing on general-purpose processors (GPPs). The early deployments of C-RAN utilized centralized BBUs composed of dedicated hardware such as ASICs (application-specific integrated circuits), FPGAs (field-programmable gate arrays) and DSPs (digital signal processors). However, this approach merely amounts to a physical separation of the analogue front-ends from the digital basebands that would be co-located in the decentralized architecture. As a result, a separate BBU is required per BS and signals need to be exchanged between the BBUs, which still makes joint processing difficult. More recent approaches [5] therefore advocate the use of more flexible GPPs, commonly found in PCs or servers. Thereby, virtualization and the concept of cloud computing as used in the IT industry [6] can be facilitated. While this further simplifies maintenance, upgrades and organization, it also introduces the economy of scale, as standardized hardware can be used, further lowering capital expenditure (CAPEX). Additionally, virtualization allows balancing processing load between the different BSs according to their load variations, thereby lowering the total processing power that needs to be deployed.

These benefits of C-RAN come at the price of a demanding transport network. In the C-RAN architecture, the transport network forwards samples between the BBUs and the RRHs, and is commonly known as the fronthaul (FH) network. In order to enable efficient, centralized and cooperative processing, the FH network must offer a huge capacity and low total latency and jitter. In fact, if not dimensioned correctly, the FH network could become a bottleneck for the performance of future networks. Furthermore, FH networks are very expensive to deploy, thus making a simple over-provisioning not economically viable. Several approaches have been investigated to reduce these FH requirements, for example by compressing FH data [7] or reducing costs by a joint optimization of the RAN and BH/FH network [8]. In this chapter, we want to describe another promising approach to address the FH challenge in future networks: a flexible centralization of RAN functionality [5, 9]. As indicated above, the current fully decentralized or fully centralized architectures are two extreme approaches. Utilizing a flexible functional split, a part of the baseband processing would be located in the BS and another in a centralized processing unit. In this chapter, we will discuss how this approach can reduce the requirements on the FH network while still partially maintaining the benefits of centralization [10]. For this, we will first introduce the considered network architecture and explain different options for a flexible centralization. We will further describe the technologies that will help to enable this approach, and describe how such a flexible split calls for a convergence of the FH and BH network to a unified transport network.

4.2   Radio Access Network Architecture

Figure 4.1 shows the architecture of a contemporary mobile network and illustrates the difference between fully decentralized, fully centralized and flexible architectures. Users (called user equipment (UE) in LTE) are located on the edge of the network and ultimately want to communicate with the core on the other edge of the network. The core then routes the traffic to the destination via gateways, and is also responsible for network management and policy control. To reach the core, the UE communicates with a BS via a RAN link. In a fully decentralized network, the BS performs baseband processing on the PHY layer and MAC layer as well as packet data convergence (PDCP, packet data convergence protocol). The traffic can then be forwarded as IP packets over BH links to the core. In the fully centralized architecture,

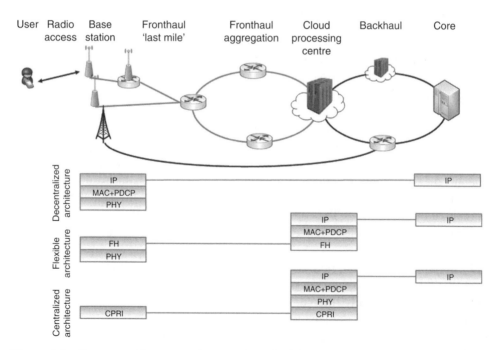

Figure 4.1   System architecture and protocol stack of decentralized, flexible and centralized networks

by contrast, all PHY, MAC and PDCP processing is performed not at the BS but in a cloud processing centre, which is connected to the core via BH. The cloud processing centre exchanges digital samples with the BS over the FH, usually using the Common Public Radio Interface (CPRI) [11] standard. The FH can be separated into an aggregation network that collects and distributes data to many BSs, and the so-called 'last mile' that represents the final link to the BSs. For the FH as well as for the BH, a number of topologies (star, chain, ring and tree) can be used [12].

The two currently used concepts of either fully decentralized or fully centralized baseband processing are the two extreme sides of a basic trade-off. The more processing is centralized, the easier it is to implement cooperative processing, the higher are the potential gains in cost savings, and the easier are maintenance and upgrading. This is traded off with a demanding and expensive FH network. The decentralized approach, on the other hand, requires a much simpler BH network but makes the aforementioned operations much more difficult or expensive. For future mobile network architectures, it therefore makes sense to look at intermediate options that still offer high centralization gains but at reduced FH requirements. In the next section, four of these intermediate split options that offer a large reduction in FH requirements will be described.

4.3   Functional Split Options

Figure 4.2 shows the more detailed PHY layer signal processing chain in the BS of a typical mobile network, using the example of LTE [13]. In the downlink (DL), the MAC layer user data are first encoded for forward error correction (FEC) before being modulated and precoded. These operations are performed according to the current channel quality and channel state, which have to be made available by measurements. Next, the user and control data are mapped to the physical resources, for example, subcarriers and time slots, thereby multiplexing different logical channels, for example, control and data channels. Additional signals for synchronization and channel measurements are added at this stage as well. In LTE, this resource mapping is performed in the frequency domain. After transforming the signal to the time domain, a cyclic prefix (CP) is added and the data are digitally filtered. Finally, the data are D/A converted, up-converted to the carrier frequency and then transmitted via the antennas.

In the uplink (UL), the process is reversed. The radio frequency signal received from the UEs is first down-converted to baseband, then digitized by sampling and quantization and digitally filtered. After the CP has been removed, the data are transformed to the frequency domain, where the different channels are demapped from the physical resources and the signals are equalized. After converting the signals back to the time domain, the symbols are detected and decoded. The resulting MAC data are then forwarded to the higher layers. In this work, we will view the hybrid automatic

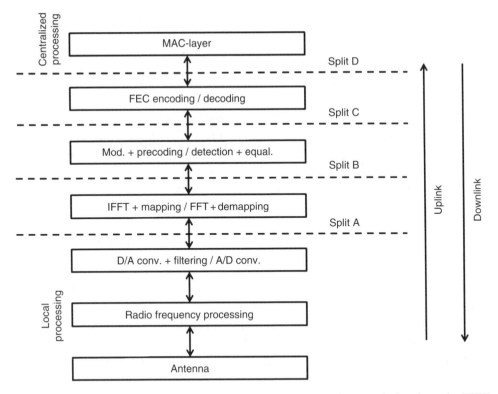

Figure 4.2   Functional split options. Reproduced from [10] with permission from the IEEE

repeat request (HARQ) as part of the decoding process, although it is often seen as part of the MAC layer.

Within this processing chain, four options of splitting the processing between the BS and the central processing unit are very promising. These are indicated in Figure 4.2 as split options A–D.

Split option A corresponds to the split used in C-RAN as implemented today. The FH interface for this split is standardized in CPRI [11]. In the DL, the complete baseband processing is performed centrally and digital samples are forwarded to the BSs. In the UL, conversely, the received signals are only digitized, filtered and then forwarded. As all baseband processing is centralized, there is no drawback in terms of what types of joint processing can be performed.

In split option B, the mapping/demapping is decentralized. In the DL, this means that the user and control data are forwarded separately and in the frequency domain. Synchronization and reference signals can be either forwarded from the central processing unit to the BSs or they may be generated and added at the BS. The different signals are combined at the BS, converted to the time domain and the CP is added before it is D/A converted. In the UL, the CP is removed, the signals are converted to the frequency domain and the separate channels are demapped at the BS.

Synchronization and channel estimation can either be performed at the BS or centrally. Due to the fact that the decentralized processing mainly involves transforming and restructuring the received signal, all types of cooperative processing can still be performed centrally without any disadvantage. However, for split B and all subsequent splits, digital processing resources are required at the BS, which makes them bigger and more power-hungry.

For split option C, modulation and precoding are also decentralized in the DL. Accordingly, bit data instead of complex amplitude samples need to be forwarded from the central processing unit to the BS. In the UL, equalization and demodulation are performed at the BSs so that soft information on user bits, for example, log-likelihood ratios (LLRs), needs to be forwarded to the central processing unit. As detection is now performed decentrally, techniques like CoMP or MPTD can no longer be performed easily, so this split does not offer as many centralization gains as the previous ones. Still, techniques like, for example, joint decoding [14] can be performed centrally. Also, techniques such as in-network processing have been proposed [15] that enable a decentralized, yet joint detection. However, these methods require extensive inter-BS links to exchange signals, thereby again increasing the requirements of the FH.

In split option D, coding and decoding are also performed decentrally, that is, all PHY-layer processing is performed at the BSs. Accordingly, centralization gains can only come from higher-layer processing. Techniques like joint scheduling [16] and connection control [17] still offer benefits from such a partly centralized architecture. However, the requirements of these higher-layer splits are very similar to those of split D.

4.4   Requirements of Flexible Functional Splits

The main benefit of an only partial centralization is the reduced requirements on the FH network in terms of data rate, latency and jitter. In the following, the requirements of the four introduced split options will be discussed. While the required data rates can be relatively easily derived, the requirements on latency and jitter are difficult to estimate, as they depend on numerous physical parameters as well as standardization and implementation aspects. The data rates in this section are derived for a single sector and a single carrier/frequency band. If a site incorporates multiple sectors and carriers – as is quite common in today's network – the data rates scale linearly with that number. This is sometimes identified as the number of 'sector-carriers'. Furthermore, LTE is used as a baseline technology, as it has the most demanding requirements when compared to third generation (3G) and second generation (2G) technologies. However, the derivations of other previous or future standards are expected to be very similar. Some numerical examples for contemporary and future networks are given at the end of this section.

4.4.1   Split A

Split A marks the limit between analogue and digital signals. Accordingly, the data exchanged via the FH correspond to digitized I/Q samples and the data rate required for split option A in bit/s can be calculated as follows:

$$D_A = N_A \cdot f_s \cdot N_{Q,A} \cdot 2 \cdot \gamma, \tag{4.1}$$

where N_A is the number of antennas, f_s is the sampling frequency and $N_{Q,A}$ is the resolution of the quantizer in bits. The factor 2 accounts for the I- and Q- phase of the signal and γ is the overhead introduced by the FH, that is the overhead of line coding or an FEC, or additional control signals. The number of antennas N_A indicates that a separate data stream needs to be fronthauled for each antenna, while the sampling frequency f_s accounts for the system's bandwidth. The quantizer resolution $N_{Q,A}$ and the factor 2 account for the word length per digital sample.

The dependence on the number of antennas N_A of this split may become critical in future mobile networks, when massive multiple-input/multiple-output (MIMO) techniques [18] with 100 or more antenna elements may be introduced, as this linearly scales up the FH data rate. Similarly, f_s depends on the total bandwidth, which is also foreseen to increase in future networks. The quantizer resolution needs to be quite high, usually around 15 bits per dimension, due to the high dynamics of the time domain signal in LTE [19]. However, the main disadvantage of this split is the fact that it does not depend on the actual user traffic, that is even when no user is connected to the BS, the full FH data rate needs to be forwarded.

In terms of latency, CPRI defines a maximum round-trip time of 5 microseconds (µs) for processing, which is added to the propagation time to amount, typically, to a few hundred µs. According to the standard, this is mainly motivated by the inner loop power control for the UL in UTRAN (universal terrestrial radio access network, the mobile network of 3G) [11]. To compensate for the fast fading and near–far effects, the latency must be in the order of the coherence time of the channel, which can be as low as one millisecond (ms) for fast users in current 4G systems. As the coherence time decreases with higher carrier frequencies, the requirements can be expected to be even stricter in 5G systems which utilize millimetre-wave (mmWave) frequencies. Other works [19, 20] state that the time allowed in LTE for HARQ acknowledgements is the most critical latency constraint. The UE expects an acknowledgment after 3 ms. When subtracting the time typically required for baseband processing, this leaves a few hundred ms for the FH. However, this is a direct result of the LTE standard and not of physical constraints, and thus could be modified in future standards. For split C, we describe methods for mitigating the delay requirement of HARQ.

In order to enable CoMP and distributed MIMO techniques, samples of different BSs need to be correctly aligned and the timing advance, that is, the different propagation times of different UEs, need to be known precisely. For this, the total delay on the FH must be measured and jitter must not be too high. Hence, the accuracy of the timing measurements and the jitter need to be in the order of magnitude of the sample duration, that is, a few tens of nanoseconds (ns) for LTE. In future networks utilizing bandwidths in the GHz range, these requirements will increase still further as the sample duration decreases. This could lead to a maximum tolerable jitter in the order of hundreds of picoseconds.

4.4.2 Split B

The data exchanged when utilizing split option B corresponds to frequency domain samples. Additionally, the different physical channels are separately available. These different channels carry data responsible for, for example, synchronization, channel estimation, control signals or user data. Some of these channels, for example, synchronization and reference symbols, do not have to be fronthauled as they can be generated at the BSs. These signals do not benefit from centralized processing and can be calculated based on a few parameters like the cell identifier, which can be assigned permanently. The required data rate for split B can be calculated as:

$$D_{\mathrm{B}} = N_{\mathrm{A}} \cdot N_{\mathrm{SC}} \cdot N_{\mathrm{S}} \cdot T_{\mathrm{F}}^{-1} \cdot \eta \cdot N_{\mathrm{Q,B}} \cdot 2 \cdot \gamma, \tag{4.2}$$

where N_{SC} is the number of utilized subcarriers, N_{S} is the number of symbols per frame, T_{F} is the frame duration, η the percentage of actually occupied resources and $N_{\mathrm{Q,B}}$ the number of quantization bits per sample for this split. N_{SC} points to the fact that usually guard carriers are used to avoid interference with neighbouring bands. As they do not carry any data, they can be added after/discarded during the mapping/demapping process. To give an example, a 20 MHz LTE system has 2048 subcarriers in total, of which 848 are used as guard bands, leading to $N_{\mathrm{SC}} = 1200$ utilized subcarriers. Similarly, N_{S} is the number of symbols that carry user or control channels, as synchronization and reference signals also potentially do not have to be forwarded between the BS and the central processing unit. The load factor η accounts for the fact that before/after mapping/demapping only those physical resources need to be forwarded that actually carry user data, that is, if the BS is only 50% loaded, the FH data will also be reduced by a factor of 2. As the dynamics of the frequency domain signal are much lower compared to time domain signals, the number of bits per sample $N_{\mathrm{Q,B}}$ can be reduced to 7–9 bits [19], thereby further reducing the FH data rate.

Split B offers the option of performing channel estimation decentrally, as the reference symbols are available after demapping. This would remove the power control constraints on the latency for this split, as the signal-to-interference-plus-noise ratio (SINR) could be calculated in the BS. However, the centralized precoding still requires that up-to-date channel information is available at the central processing unit. As a consequence, the latency requirements could be slightly relaxed, depending on the coherence time of the channel. Jitter is less critical for the higher-layer splits, as the samples are aligned at the BS. Still, processing in the central unit can only start once the symbols from all BSs are available. While the symbols can be buffered to compensate for different latencies, a high difference in arrival times of different BSs would lead to a large latency.

4.4.3 Split C

Split C is the only split that exhibits a significant asymmetry between UL and DL. In the DL, encoded user bits are transported, while in the UL, LLR values have to be forwarded to enable turbo decoding. Still, the required data rate for both UL and DL can similarly be calculated as:

$$D_{\mathrm{C}} = N_{\mathrm{L}} \cdot N_{\mathrm{SC}} \cdot N_{\mathrm{S}} \cdot T_{\mathrm{F}}^{-1} \cdot \eta \cdot \log_2 M \cdot N_{\mathrm{Q,C}} \cdot \gamma, \tag{4.3}$$

where N_{L} is the number of layers (spatial streams), M is the modulation order and $N_{\mathrm{Q,C}}$ is the number of quantization bits for split C.

The difference between UL and DL lies in $N_{\mathrm{Q,C}}$. In the DL, only encoded user bits are forwarded, so the number of bits per symbol is coupled to the utilized modulation scheme, that is 2 bits for 4-QAM, 4 for 16-QAM and so on. In the UL, one LLR per information bit has to be forwarded and each LLR is typically represented by 3 bits, that is, 6 bits for 4-QAM, 12 bits for 16-QAM and so on.

The main advantage of this split is that the number of antennas is mapped to spatial streams and vice versa. To give an example, if a BS is equipped with four antennas but due to the channel state can only transmit one spatial stream to a user, only this one stream has to be forwarded instead of one stream for each of the four antennas. This will become of importance in massive MIMO systems [18], when a high number of antennas is mostly used for beamforming, and only a limited number of independent user streams are transmitted. Both the number of layers and the modulation scheme depend on the current channel quality of a certain user. These two dependencies result in the fact that the FH traffic is even more coupled with the actual user traffic, that is, when a user faces bad channel conditions and can transmit only little data, this is reflected in the FH traffic.

As channel decoding is still performed centrally, the HARQ scheme of LTE can be limiting, as UEs require an acknowledgement to be sent within 3 ms, so split C needs to meet that latency requirement. To overcome this, local feedback schemes have been proposed [21] in which the BSs send feedback without waiting for confirmation about the decoding outcome from the central processing unit. However, this will reduce the performance to some degree. Alternatively, it has been proposed to suspend the HARQ process if the acknowledgement cannot be sent in time and simply not schedule new traffic until the decoding is finished [19]. For future systems, increasing the allowed HARQ delay could also be considered. However, this requires higher available memory resources at the UEs, as the data of the different HARQ processes will have to be kept until the acknowledgement arrives.

4.4.4   Split D

Using split D, bit-level user data are transported and thus correspond very closely to what is considered classical BH in current networks. The required data rate can be calculated as:

$$D_{\mathrm{D}} = N_{\mathrm{L}} \cdot N_{\mathrm{SC}} \cdot N_{\mathrm{S}} \cdot T_{\mathrm{F}}^{-1} \cdot \eta \cdot \log_2 M \cdot R_{\mathrm{c}} \cdot N_{\mathrm{Q,D}} \cdot \gamma, \tag{4.4}$$

where R_{c} is the code rate and $N_{\mathrm{Q,D}}$ is the number of quantization bits for this split. Coding/decoding adds/removes redundant bits to/from the actual information bits. These redundant bits do not have to be forwarded in this split, which further decreases the FH data rate. Also, the code rate is coupled to the channel quality and therefore to the actual user traffic. As information bits are forwarded, the number of quantization bits can be set to one for this split.

Split D terminates PHY-layer processing. Hence, the latency requirement is determined by the higher layers. If centralized scheduling is performed and intended to exploit time diversity, the latency must be small enough to follow the fading of the channel. Otherwise it can be relaxed to the application-layer requirements which are typically in the order of a few tens of ms.

4.4.5   Summary and Examples

Table 4.1 summarizes the main parameters that determine the requirements of the different splits.

The different data rate requirements are further illustrated in Figure 4.3. The requirements for the four different splits are depicted for four different parameter sets. The baseline is a 4G system, that is, LTE, which is further differentiated in a maximum data rate, utilizing the highest parameters possible and an exemplary parameter set using more representative values. The exemplary parameter set was chosen to illustrate additional reductions in data rate that would not be visible in the maximum data rate. To give a few examples, the maximum load of a BS is 100%, while in real deployments such a utilization would actually indicate a capacity problem of the network as a BS should be loaded less than 100%. Similarly, the maximum code rate is 1.0 (i.e., uncoded transmission), while real systems have to employ a lower code rate according to the channel quality.

As we discussed previously, the requirements are expected to increase in future networks mainly through three advancements: the introduction of massive MIMO and the utilization of higher carrier frequencies and bandwidths. Accordingly, Figure 4.3

Table 4.1 Parameters impacting FH requirements for different functional splits

Split	Data rate	Latency	Jitter
A	• Number of antennas • Bandwidth • Quantization in time domain (high number of bits)	• Channel coherence time • UL power control needs to be able to follow fast fading	• Sample duration • Timing advance needs to be measured exactly
B	• Number of antennas • Number of utilized subcarriers • Load • Quantization in frequency domain (low number of bits)	• Channel coherence time • Precoding must be able to follow channel	• Queueing should not lead to increased total latency
C	• Number of spatial layers • Load • Modulation order • Number of quantization bits 1 (DL) or 3 (UL)	• Maximum delay of HARQ acknowledgement	• Queueing should not lead to increased total latency
D	• Number of spatial layers • Load • Modulation order • Code rate	• Requirements of higher-layer applications • For joint scheduling, channel coherence time can still be determining factor	• Queueing should not lead to increased total latency

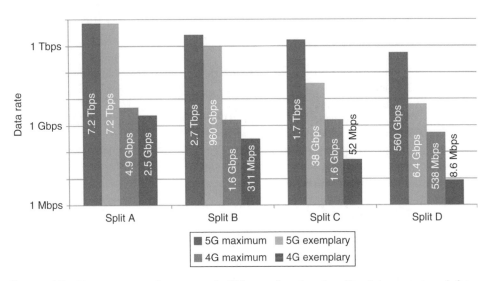

Figure 4.3 Data rate requirements of different functional splits for current and future networks

also shows data rates for a potential 5G system with 100 antennas and a sampling frequency of 1.5 GHz. As higher-order modulation schemes are also under discussion, 1024-QAM was assumed. To account for the lower symbol distance in 1024-QAM, a higher quantizer resolution was also taken into account. The full list of parameters can be found in Table 4.2. In rows where only one parameter is listed, the exemplary and maximum values are identical.

From the chosen parameters it is obvious that a potential 5G system would further increase the already demanding requirements. In particular, the scaling with number of antennas illustrates clearly that a full centralization with per-antenna backhauling is infeasible. In fact, the data rate is increased by more than three orders of magnitude. While it can be expected that transport network technologies will also advance in the future, such an increase cannot be expected in the timeframe considered for 5G mobile networks. It can be further seen from Figure 4.3 that while the higher-layer splits lead to a reduction in general, there is also an important difference between the maximum possible data rate and the exemplary rate. This difference is discussed in detail in Section 4.5.

Furthermore, Table 4.3 shows the impact of higher carrier frequencies, larger bandwidth and higher UE speeds on the channel coherence time and sample duration, which, in turn, determine the maximum tolerable delay and the delay accuracy for a fully centralized system (split A), respectively. From the forecast numbers for a 5G system, it is clear that these requirements would be even more challenging to fulfil than those of a 4G system.

Table 4.2 Parameters for the calculation of data rate requirements

Symbol	Description	5G max. / exemplary	4G max. / exemplary
N_A	Number of antennas	100 / 100	4 / 2
N_L	Number of spatial layers	50 / 8	4 / 1
N_{SC}	Number of subcarriers	60 k / 50 k	1200 / 1080
N_S	Number of data symbols per frame	14 / 12	14 / 12
f_s	Sampling frequency (bandwidth)	1.5 GHz	30.72 MHz
N_Q	Number of quantization bits per I / Q dimension for split A, B, C, D	18, 12, 3, 1 bit	15, 9, 3, 1 bit
γ	FH overhead	1.33	1.33
T_F	Frame duration	1 ms	1 ms
η	Utilization (load)	1.0 / 0.5	1.0 / 0.5
M	Modulation order	1024 / 16	64 / 4
R_c	Code rate	1.0 / 0.5	1.0 / 0.5

Table 4.3 Delay requirements of 4G and exemplary 5G systems for split A

Parameter	4G	5G
Carrier frequency f_C	2 GHz	70 GHz
Max. UE speed v	250 km/h	500 km/h
Channel coherence time/max. delay $(\approx 0.423 \cdot \dfrac{c}{v \cdot f_C} \cdot)$ [22]	**914 µs**	**13 µs**
Bandwidth	20 MHz	1 GHz
Sampling rate f_s	30.72 MHz	1.5 GHz
Sample duration/delay accuracy $(= 1/f_s)$	**32.6 ns**	**0.67 ns**

4.5   Statistical Multiplexing in a Flexibly Centralized Network

The previous section observed how a flexible split in general reduces the data rate requirements by not forwarding certain parts of the signal. One main disadvantage of split A, which corresponds to the currently used split in C-RAN, is identified with the fact that the FH data rate is always constant and does not vary with the actual user traffic. With higher-layer splits, this coupling is progressively increased, which gives rise to another important aspect of a flexible functional split: the statistical multiplexing gain.

As described in Section 4.2, FH networks typically consist of two parts: the so-called 'last mile' that connects the individual BSs and an aggregation, or 'metro', network that aggregates that traffic from numerous BSs and forwards it to the core network. The observations made in Section 4.4 are mainly valid for the last mile, as the data rates for single BSs are derived. In the aggregation network, multiples of the thus-described data streams have to be forwarded over a single link. Because the individual last mile traffic is time variant, the dimensioning of these aggregation links now poses a trade-off: on the one hand, the aggregation network has to be able to forward peak traffic, but on the other hand, peak traffic will only occur in a very limited number of times, resulting in an underutilization of the expensively deployed network for most of the time. In the following, we describe how this can be addressed to dimension the network appropriately.

4.5.1   Distribution of FH Data Rate per Base Station

The variance of the FH traffic is different for the four functional splits and depends on the respective parameters in Equations (4.1)–(4.4). The variant parameters in those equations are $N_{\mathrm{L}}, \eta, m = \log_2 M$ and R_c. While the load η varies due to the changing

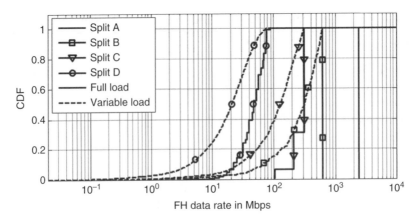

Figure 4.4 Cumulative distribution function (CDF) of the data rate for different functional split options with full load (solid lines) and an exemplary variable load (dashed lines). Reproduced from [10] with permission from the IEEE

traffic demand generated by the users, the remaining parameters depend on the channel quality of the users. In order to use a high number of layers, the channel needs to offer a high spatial separation, while the remaining parameters depend on the SINR. To illustrate the impact of those varying parameters, Figure 4.4 shows the cumulative probability function (CDF) of a single BS for the different split options. The SINR distribution was taken from system-level simulations in a dense urban scenario. As the load distribution has the largest impact, we show the CDFs both for a full load and a variable load with a uniform distribution between 0 and 1. The remaining parameters are chosen identically to the exemplary 4G system in Table 4.2.

It can be observed that the data rate for split A is constant, which is a major disadvantage. Split B only varies with the load and is therefore constant in the case of a full load. Split C additionally varies with the modulation scheme, which can be seen from the three steps in the fully loaded case, corresponding to the three modulation schemes 4-QAM, 16-QAM and 64-QAM. Similarly, split D depends on the modulation and coding schemes (MCS), so 28 steps can be observed, corresponding to the 28 utilized MCSs in this example. For the case of varying load, these distributions are additionally convoluted with the load distribution. In general, we can observe that, on the one hand, the maximum data rate in each split decreases, and that the data rate becomes more variant. This variance is exploited for statistical multiplexing.

4.5.2 Outage Rate

To avoid having to dimension the network for peak traffic, usually a certain outage probability is defined [23], that is, it is conceded that for a certain percentage of time, the network cannot transport the offered traffic. This will potentially lead to a reduced

QoE (quality of experience) for the users, but is accepted by the operators as it can reduce the required FH capacity dramatically. As an example, assume the traffic of one BS is approximately Gaussian distributed[1] with $D \sim \mathcal{N}(\mu_D, \sigma_D^2)$. Then the probability that the actual data rate D exceeds the deployed capacity D_d can be calculated as:

$$P_o = P(D > D_d) = \frac{1}{2}\text{erfc}\left(\frac{D_d - \mu_D}{\sqrt{2\sigma_D^2}}\right),\tag{4.5}$$

with erfc being the complementary error function. This probability is also known as the outage probability.

Conversely, we can calculate the capacity that has to be deployed to yield a certain outage probability as:

$$D = \text{erfc}^{-1}(2P_o)\sqrt{2\sigma_D^2} + \mu_D.\tag{4.6}$$

As this data rate is identical to percentiles of the corresponding data rate distribution, we write D_{1-P_o} to easily identify a data rate with its outage probability, that is, we write D_{99} to identify the data rate that needs to be deployed for an outage probability of 1%. We also call this data rate the outage rate.

Due to congestion effects, outages can already occur at loads below 100%. To account for this, the outage rate is usually divided by a safety factor ϵ to calculate the actually deployed data rate D_d [23]:

$$D_d = \frac{D_{1-P_o}}{\epsilon}.\tag{4.7}$$

For example, an outage probability of 1% and a safety factor $\epsilon = 0.9$ imply that the link will be loaded less than 90% for 99% of the time.

Similar observations, of course, can be made for distributions other than the Gaussian distribution used here as an example. However, the outage capacity is only relevant if the data rate follows a varying distribution. Hence, the outage capacity for the constant data rate of split A is always equal to the maximum data rate.

4.5.3 Statistical Multiplexing on Aggregation Links

The statistical multiplexing gain comes into effect in the aggregation network, when multiple streams with varying data rates are added up. The most straightforward way

---

[1] In fact, the load distribution can never be truly Gaussian, as this would allow for negative loads. However, the assumption will make the comparison in Section 4.5.3 easier.

to dimension the aggregation link would be simply to scale the data rate for one BS with the number of BSs being the aggregate, for example, if N BSs are aggregated:

$$D_{\text{d,aggr,nomux}} = ND_{\text{d}} = \text{erfc}^{-1}(2P_{\text{o}})N\sqrt{2\sigma_D^2} + N\mu_{\text{D}}.$$ (4.8)

However, this neglects the statistical multiplexing. The aggregation of several varying data streams can be seen as a summation of random variables. From the central limit theorem [24], it is known that the distribution of the sum of random variables will converge to a Gaussian distribution. If all data rates follow the same distribution, the sum will, in fact, be distributed as $D_{\text{d,aggr}} \sim \mathcal{N}(N\mu_{\text{D}}, N\sigma_D^2)$. Accordingly, the outage probability can now be calculated as:

$$P_{\text{o,aggr}} = P\left(\sum D_i > D_{\text{o,aggr}}\right) = \frac{1}{2}\text{erfc}\left(\frac{D_{\text{o,aggr}} - N\mu_{\text{D}}}{\sqrt{2N\sigma_D^2}}\right).$$ (4.9)

and the outage rate as:

$$D_{\text{d,aggr}} = \text{erfc}^{-1}(2P_{\text{o}})\sqrt{N}\sqrt{2\sigma_D^2} + N\mu_{\text{D}}.$$ (4.10)

The convergence of probability density functions (PDFs) to a Gaussian distribution is illustrated in Figure 4.5, using split B as an example. The load is uniformly distributed between 0 and 1, which yields a rectangular PDF for a single BS. When adding multiple varying data streams, the corresponding PDFs are convoluted. This yields a triangular distribution for two BSs and increasingly Gaussian-like distributions for four and eight BSs, as depicted. Figure 4.5 additionally illustrates the rationale behind

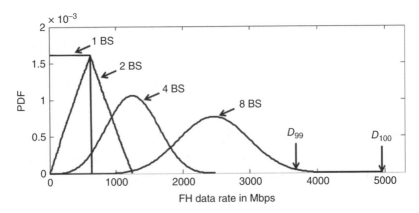

Figure 4.5 PDFs of aggregated FH traffic of one, two, four and eight BSs, and data rate percentiles

the outage rates. While the maximum data rate D_{100} for eight BSs is close to 5 Gbps, the 1% outage rate D_{99} is approximately 3.6 Gbps. Consequently, about 1.4 Gbps of aggregation capacity can be saved by accepting a mere 1% of outage probability.

As can be seen from Equation (4.10), the aggregated outage rate scales with $\sqrt{N}$ of the standard deviation, while the outage rate in Equation (4.8) scales with N times the standard deviation. This difference corresponds to the statistical multiplexing gain. Practically, this gain occurs because it is very unlikely that a large number of streams will carry peak traffic at the same time, hence the aggregated probability distribution is flattened out. The resulting multiplexing gain can be calculated as:

$$g_{mux} = \frac{D_{d,aggr,nomux}}{D_{d,aggr}} = \frac{\sqrt{N}\left(\alpha\dfrac{\sigma_D}{\mu_D}+1\right)}{\alpha\dfrac{\sigma_D}{\mu_D}+\sqrt{N}}, \tag{4.11}$$

with $\alpha = \sqrt{2}\mathrm{erfc}^{-1}(2P_o)$. For a large number of BSs this converges to:

$$\lim_{N\to\infty} g_{mux} = \alpha\frac{\sigma_D}{\mu_D}+1. \tag{4.12}$$

Figure 4.6 illustrates this multiplexing gain using the data rate distribution from Section 4.5.1. The x-axis shows the deployed FH data rate $D_{d,aggr}/\epsilon$ for $P_o = 99\%$ and $\epsilon = 0.9$ and the y-axis the number of BSs that could be supported with a single aggregation link of corresponding capacity.

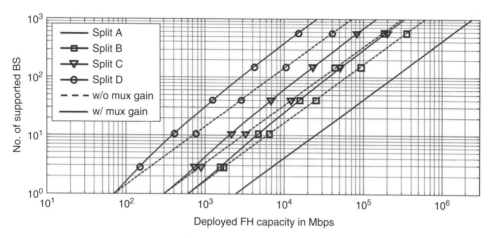

Figure 4.6 Number of supported BSs versus deployed FH capacity of different functional split options with and without multiplexing gain (solid and dashed lines, respectively). Reproduced from [10] with permission from the IEEE

For split A there is no aggregation gain because the FH data rate is constant. For the higher splits, a statistical multiplexing gain of up to a factor of 3 can be observed, with the gain being more pronounced for a larger number of BSs and for higher-layer splits. This can again be explained with probabilistic methods.

Equations (4.1)–(4.4) contain four factors that are varying: N_L, η, m and R_c. The modulation scheme and code rate are typically chosen together and are therefore not independent. We therefore define the spectral efficiency:

$$s = m \cdot R_c \tag{4.13}$$

Otherwise we can expect that N_L, η and s are independent. With this, the mean of the distribution of the product of these variables can be calculated as:

$$\mu_D = \mu_{N_L} \mu_\eta \mu_s, \tag{4.14}$$

and the variance is given by:

$$\sigma_D^2 = \left(\sigma_{N_L}^2 + \mu_{N_L}^2 \right)\left(\sigma_\eta^2 + \mu_\eta^2 \right)\left(\sigma_s^2 + \mu_s^2 \right) - \mu_{N_L}^2 \mu_\eta^2 \mu_s^2. \tag{4.15}$$

From Equation (4.12) and Equation (4.15) we now see that the total multiplexing gain depends on the ratio of σ_D and μ_D, which, in turn, depend on the means and variances of the parameters N_L, η, m and R_c. In summary, it can be said that the statistical multiplexing gain is larger the more the individual parameters vary. An illustration for this is given in Figure 4.7. It shows the number of supported BSs versus the deployed FH capacity for split D, but for different load distributions. The load distributions are uniformly distributed between 0 and 1, 0.2 and 0.8, 0.4 and 0.6, and constant with $\eta = 0.5$, respectively. This yields four distributions with the same

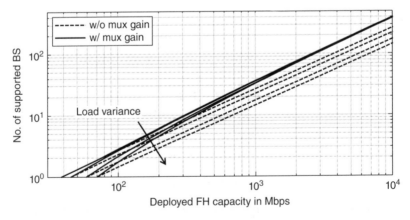

Figure 4.7 Number of supported BSs versus deployed FH capacity of split option D for uniform load distributions with different variances

mean load $\mu_\eta = 0.5$ but different variances of 0.0833, 0.0300, 0.0033 and 0, respectively. As can be seen, the multiplexing gain is higher for larger variances, as predicted by Equation (4.12). In other words, the over-provisioning of not considering the multiplexing gain is much worse if the traffic is varying heavily.

4.6 Convergence of Fronthaul and Backhaul Technologies

The concept of the flexible functional split describes a gradual variation between what is currently known as FH and BH. Due to the very different requirements of these two 'extreme' versions of a split, they are, in fact, deployed as two different segments of the network. This means that not only is separate hardware deployed but also that completely different and largely incompatible standards have evolved. While this has to be owing to historic development, it is a highly suboptimal solution, as it not only drives up costs but also makes the management of the network more complex. Even more importantly, the complexity would further increase if the same approach was followed for a flexible functional split, that is, developing different hardware and standards for each of the split options. Instead, it is highly desirable to converge the technologies of the current two options and design a unified transport network. This unified transport network should not only be able to support any number of different functional splits but should also be agnostic with respect to the underlying access technology to make it future proof for the coming generations of mobile networks. In the following, an overview of the existing solutions across different layers is given and convergence to a unified technology is discussed.

4.6.1 Physical Layer Technologies

PHY-layer technologies pose an upper bound on the performance that can be achieved in FH networks. Additionally, they have a large impact on the cost of the FH links, as specific hardware is required for each technology. In general, PHY-layer technologies can be grouped into wired and wireless technologies. The wired technologies can be further divided into fibre/optical technologies and copper/electrical technologies, while wireless technologies are usually distinguished by the carrier frequency they utilize.

The most intuitive – and hence most widely used – option in current C-RAN deployments is fibre. It offers very large capacities of tens of Gbps, low latency and high reliability. The range is only limited by the tolerable latency and the cost of repeaters. Dedicated point-to-point links have been used so far for FH in fully centralized C-RAN, meaning that one fibre core or one wavelength channel is dedicated to an FH link from RRH to BBU. In order to exploit statistical multiplexing and to avoid a heavy underutilization of the fibre in a flexible C-RAN deployment, time-shared optical networks could be used [25] in the future. The main disadvantage of fibre technology is that it is expensive and slow to deploy due to the extensive civil works required. Adding right-of-way issues to that prohibits fibre connectivity to every BS, especially in the case of dense deployments of small cells. While some operators have their own fibre networks, others – especially non-incumbent operators – have to rely on expensive third-party leases, further complicating the business case.

In the context of classical BH, that is, very high-layer splits, a number of other PHY technologies are in use, making them principal candidates for a flexible C-RAN solution. Copper-based solutions like coax cables face the same economic problems as fibre but in addition offer less capacity. In some cases, existing deployments can be utilized to mitigate that effect. For example, digital subscriber line (DSL) connection of shop owners could be used to provide FH for indoor BSs if a corresponding incentive was offered. However, due to the limited capacity and protocol overhead of DSL technology, even with new technology generations such as G.fast [26], this is only an option for higher-layer splits.

Wireless options, in general, offer the benefit that they are cheaper and faster to deploy, as only comparatively little civil work is required. On the other hand, their range and reliability are limited by the much higher path loss.

One category of wireless transport solutions uses similar carrier frequencies to the access links, that is, sub-6 GHz bands. These are more suitable for high-layer splits, as their respective capacities are comparable with the access link technology. The latency can also be quite high because similar protocols to the access link have to be observed. While some of these technologies can, in principle, use the same hardware as the access link and therefore benefit from economies of scale, additional licensed frequency bands have to be acquired or already available ones have to be assigned, which again makes the overall deployment quite expensive. On the other hand, sub-6 GHz technologies offer point-to-multipoint connections, which can significantly reduce the number of required antennas and thus reduce CAPEX.

Microwave technology up to a few tens of GHz is already widely used in classical BH and could also be an option for lower-layer splits. With capacities of currently up to 1 Gbps and a range of a few tens of kilometres, it can be utilized for the 'last mile', although not for the aggregation network. The main disadvantage is that, again, licensed frequency bands are utilized, which increases cost and deployment time. Also, the limited range requires the set-up of potentially multiple solid masts for the intermediate antenna dishes along the transmission path. This results in the need for real estate or roof access, all of which comes with administrative and cost-related overhead.

Millimetre-wave technology, also sometimes referred to as V-band or E-band technology depending on the corresponding frequency bands between 60 and 90 GHz, is another good option for wireless FH. While still a relatively new technology, it already offers a capacity of up to 10 Gbps and per-hop latencies as low as

10 ns. In addition, it utilizes unlicensed or lightly licensed bands, which is advantageous in terms of cost and deployment time. However, mmWave frequencies suffer from relatively high free-space path loss. Although this can be partially compensated with high-gain antennas, it still limits the total range to a few kilometres at most for E-band and a few hundred metres for V-band. With current research into large-scale antenna arrays with hundreds of individual antenna elements, beamforming techniques could be used to enable a flexible selection of point-to-point links. Thereby, time sharing or quick reconfiguration of interconnections in cases of large changes in traffic distribution could be enabled. Due to the low wavelength at mmWave frequencies (5 mm at 60 GHz), the corresponding antenna arrays can be built relatively small.

Finally, free space optics (FSO) is a wireless technology that utilizes laser light for communication. While it offers similar capacities and latencies to mmWave technology, it is susceptible to misalignment and wind sway due to its narrow beam width. Weather effects like fog or snow can also limit the range or decrease the reliability. Still, it is a viable option for the last mile in scenarios where these effects can be neglected.

In summary, a more flexible approach to the functional split can benefit from a more heterogeneous selection of FH technologies. Lower cost and higher deployment speed make wireless technologies attractive on the last mile and the first level of aggregation. Range limitation can be overcome by multi-hop chains at the cost of increased latency. Still, it is undisputed that the higher levels of the aggregation network have to be composed of fibre, which alone offers the capacity to aggregate hundreds of BSs.

To further illustrate the suitability of the discussed technologies for the different functional splits, Figure 4.8 shows the requirements of the four splits A–D in terms of latency and capacity (for a single BS) versus the different technologies, based on a market study in [27]. As can be seen, the utilization of split B or C would already enable a more heterogeneous selection of applicable PHY technologies. It should be noted that all discussed FH technologies can be expected to advance in the future; however, access link technology will similarly progress, so that the mapping can still be considered viable for future networks.

4.6.2 Data/MAC Layer Technologies

On the MAC layer, there currently exist two completely different standards corresponding to the two different extremes of the functional split: full centralization and full decentralization. In fully centralized C-RAN architectures, CPRI has evolved as the *de facto* industry standard for digital FH, while the classical BH in decentralized networks utilizes Ethernet. In terms of the data plane protocol, the incompatibility of these two standards arises mainly from the different frame formats that are used and

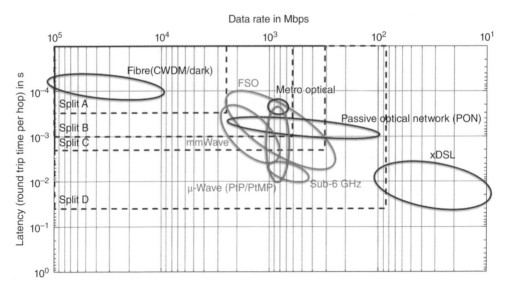

Figure 4.8 Functional split requirements versus physical layer technologies. Reproduced from [10] with permission from the IEEE

the strict requirements on timing and jitter that were introduced for CPRI. This incompatibility has resulted in the fact that two separate transport networks have to be deployed, which is a suboptimal solution as it increases deployment cost and management overhead.

At first glance, the introduction of a flexible split seems to further complicate this issue, as neither of the two existing standards will be compatible with all intermediate splits. However, this offers the chance to design a unified transport technology that is not only compatible with the existing two standards, but with a wide range of functional splits. Such a unified technology does not need to be designed from scratch but could be based on one of the existing standards. Investigations are already under way to enable CPRI over Ethernet [28], while also taking the possibility into account of mapping arbitrary data to Ethernet frames. The main challenge in such a unified frame format is to meet the strict requirements of the lower splits in terms of latency and jitter. To achieve them, large improvements in time synchronization as well as a reduction in deterministic delays will have to be achieved. GPS-assisted synchronization or synchronous Ethernet using the precision time protocol [29] are available and could be utilized for this. In terms of capacity, 10 Gbps Ethernet is already standardized (802.3.ae, ak, an [30]) and study groups are investigating data rates up to 400 Gbps [31]. On the other hand, a recent initiative aims to standardize a more flexible FH in the form of the next generation fronthaul interface (NGFI) [20]. However, while its goal is to make the FH more flexible and support higher-layer splits, it is currently not considering utilizing it for BH as well.

4.6.3   Network Layer Technologies

The network layer is responsible for switching and routing traffic between network nodes. As such, it has a decisive impact on a converged transport network architecture. While BH networks inherently work with packet switching, FH networks commonly use dedicated point-to-point connections. A converged transport network will have to support packet switching and point-to-multipoint connections, as this simplifies implementation and management and allows for arbitrary placement of the centralized processing elements, as discussed in Section 4.7. Currently, they can be located only at sites with sufficient dedicated links to support the RRHs, leading to a concentration in a low number of large data centres. However, to enable the fast migration of RAN functionality from one data centre to another, and to adapt to varying load profiles across BSs, packet switching is the only efficient option.

One challenge for this is that additional delays are introduced due to the processing time of switches and routers. Especially for very large networks with many hops, this could make low-layer functional splits infeasible. However, recent advances in low-latency switching have decreased the switching delay to about 100 ns, which is three orders of magnitude lower than the requirements for a fully centralized network.

A more important impact comes from queueing, as this not only introduces additional delay but it is also indeterminate as the length of the queue will vary with the variation in packet arrival. Prioritization of packets can be used to reduce the total wait time for certain packets. This is managed via quality of service (QoS) class identifiers; however, the means of choice for providing fairness among QoS classes are probabilistic scheduling methods, adding another layer of unpredictability. For a converged transport network architecture, new methods would have to be found to make the delay more predictable and guarantee maximum latencies. Special QoS classes for RAN functionalities could be introduced that prioritize corresponding packets above everything else. For times of congestion, the dropping of other packets could be considered. The requirements for higher-layer applications are, in fact, many orders of magnitude larger than RAN packages, for example, Voice-over-IP requires latencies of tens of ms as compared to the hundreds of µs required for a fully centralized RAN architecture. Hence, RAN requirements instead of application-layer requirements will dominate the design of a converged BH/FH network. On the other hand, the application layer could also benefit from greatly reduced latencies, enabling 'tactile Internet' applications [32] like virtual reality, automation control or remote vehicle steering. Still, this would require an update of the whole BH network. While the queueing policy can usually be modified via software or firmware updates, low-latency switches require new hardware at substantial cost. Also, switches need to enable network synchronization as discussed in the section above, which is not widely supported by the currently deployed hardware.

4.6.4   Control and Management Plane

One common requirement from the sections above is that future networks need to be more flexible. The introduction of more heterogeneous PHY technologies, a unified frame format able to carry traffic from different splits and packet-switched FH all support a transport network that can be adapted to follow, for example, hourly or daily traffic variations. Still, the evolution to 5G networks will bring along a further process of updating existing deployments or adding more hardware or functionalities. In particular, with the advent of mmWave access technology, massive MIMO and the tactile Internet, the requirements on the transport network will change and potentially become even more demanding. In order to adapt the network both in the short and in the long term, a universal management plane is required. This management plane will be in charge of policies ranging from connection control on the RAN and FH links over congestion control and load balancing to routing and QoS policies.

The centralization of RAN functionality also poses new challenges in terms of reliability, as the data centres hosting such functionalities are potential single points of failure, for example in the case of localized power outage or malicious attacks from hackers. As such, the control and management plane also has to be in charge of outage protection and failover control. This will include functionalities ranging from rerouting traffic in the case of a single link outage to migrating complete centralized RAN functionality from one data centre to another in the case of more severe malfunctions. In order to provide low failover times, $1 + 1$ protection schemes could be used on the transport network at the cost of additional deployed capacities.

For the implementation of such a universal control and management plane, software-defined networking (SDN) [33] and network function virtualization (NFV) [34] are good candidate technologies as they provide – among other things – unification via abstraction layers. The main challenge arises from the scale to handle a complete mobile network, including BSs, the transport networks, centralized RAN functionality and a possibly virtualized network core.

4.7   Enablers of a Flexible Functional Split

The utilization of a flexible functional split impacts not only the FH network but needs to be supported by the BSs and the central processing entity. The increased flexibility that is offered by a larger number of split options comes at the price of higher hardware complexity at the respective endpoints. In this section, a short overview is given of the technologies that are needed from the baseband processing perspective to enable a flexible functional split.

As the functional split determines the type of joint processing that can be performed, the split has to be adapted to the current scenario. For example, in a dense deployment with a large number of cell-edge users, CoMP techniques requiring a lower-layer split might yield large gains in user throughput. On the other hand, macro-cell deployments might not benefit from a high degree of centralization and therefore a higher-layer split could be employed, reducing the load on the aggregation network. This means that the functional split will have to be adapted in space and also in time to fit the scenario. This not only requires more flexible interfaces on the transport network, as discussed in the previous section, but also that the hardware both in the BS and in the central entity can perform the respective processing. This brings two problems: first, in order to support higher-layer splits, BSs would need to be equipped with the same hardware as fully decentralized BSs. As one of the intended advantages of a centralized RAN is smaller BSs, this effect would be nullified. Second, baseband processing is usually implemented on dedicated hardware like FPGAs or ASICs, which cannot be flexibly reconfigured to match any functional split. As a consequence, a more flexible hardware architecture is required in addition to the converged transport network. Recent progress in the IT industry provides very promising solutions for this with the approach of RAN virtualization and 'cloudification'.

Although the term 'cloud' was part of even the first considerations on C-RAN [1], the first deployments did not follow the principles of cloud computing as they are known in the IT industry, namely providing a 'shared pool of configurable computing resources that can be rapidly provisioned' [35]. Instead, the first deployments followed the approach of simply splitting up the conventional base stations by separating the baseband hardware from the RF front-ends [36]. However, this simple hardware centralization does not embrace the concept of cloud computing. Many advantages of centralized processing, like load balancing and energy savings, are only possible if the baseband hardware can be dynamically allocated and configured according to demand. This requires a virtualization of the available hardware, which, in general, requires the utilization of GPPs. Only if the hardware can be quickly and easily reconfigured to perform smaller or larger parts of the baseband processing can a flexible functional split efficiently be employed. The main challenge with this is that the implementation of baseband processing requires far higher throughput and lower latencies than traditional IT applications, which is why dedicated hardware is used in conventional BBUs. Additionally, the virtualization of the hardware introduces extra overhead in the form of a so-called hypervisor that oversees and controls the provisioning and operation of the virtual machines.

In order to reduce the overhead of virtualization, newer approaches to virtualized hardware employ so-called 'bare metal' servers. These do not require an additional operating system for each virtual machine. Instead, the hypervisor can directly communicate with the physical hardware. This also has the advantage that the physical hardware can be fully dedicated to a certain task, for example, a physical processor core can be assigned to the processing of a specific BS. This makes it far easier to guarantee performance when compared to a conventional virtualized system, where a physical core might have to process several virtual machines. Some recent works show that it is possible to implement computationally complex processing like turbo decoding on general-purpose hardware in real time [37]. Additionally, the amount of required processor cores can be predicted, allowing for precise provisioning in order to avoid outage by under-provisioning or wasting resources by over-provisioning [38]. However, the jury is still out as to whether the implementation of the full baseband stack on GPP is cost- and energy-efficient.

As an alternative, certain parts of the baseband processing could be outsourced to dedicated hardware accelerators while maintaining the virtualization approach. The newest generation of servers can not only be equipped with GPPs but can also be supplemented with DSP cores [39]. The operation of fast Fourier transformation (FFT) and inverse FFT (IFFT) and channel coding/decoding, in particular, can be much more efficiently performed by dedicated hardware. Field trials have already shown the validity of such a hardware-accelerator-assisted GPP implementation [40]. While these approaches are mainly designed for the central processing entity, a similar architecture can be employed on the BS side. The deployment of micro-servers or 'cloudlets' at BSs has already been envisioned, albeit mainly for the purpose of application processing [41]. However, this perfectly complements the approach of the flexible split. GPP cloudlets with additional hardware accelerators could be deployed and used for baseband processing when a higher-layer split is configured, and when a lower-layer split is configured, the now idle hardware could be utilized for user-application processing, thereby avoiding underutilization of the hardware at the BSs.

In summary, the flexibility offered by a virtualized cloud-RAN implementation will also be the basis for the flexible functional split that can be dynamically adapted in time and space to optimally reflect the scenario in terms of traffic density, fronthaul load and hardware utilization.

4.8   Summary

Taking all of the above into account, it becomes clear that the design of the FH network will be a major challenge in future networks. While the approach of C-RAN offers tremendous advantages, the FH could become a major bottleneck, both in terms of performance and cost efficiency. A more flexible functional split will help to mitigate this problem. A partial centralization can reduce the requirements on the FH dramatically in scenarios where full centralization offers no advantage. In particular, the data rate can be easily lowered and – more importantly – coupled to the actual user traffic. Network operators, therefore, need to carefully decide whether the traffic demand justifies the expensive FH for high centralization and how much gain can be expected from joint processing. In the aggregation network, the multiplexing gain plays an additional important role. As it makes use of the temporal and special traffic variation within a network, it is beneficial to aggregate traffic with a highly variable distribution. As described in Section 4.5, the variance mainly results from variable BS load and a variable SINR. Therefore, base stations exhibiting different load patterns should be aggregated, for example rural areas and inner cities, and similarly those with different SNR distributions, for example macro cells and dense small cells. The effect will be even more pronounced in a converged BH/FH network. By adding further traffic in the form of BH and control signalling, the traffic becomes more diverse, thereby increasing the benefit of statistical multiplexing.

The main benefit of the reduced FH requirements will come in the form of reduced deployment costs. Less FH capacity per BS means that less capacity needs to be deployed. To summarize the effect of statistical multiplexing and the general data rate reduction of higher-layer splits on the deployment cost, Figure 4.9 illustrates the number of BSs that can be supported by a single link using exemplary FH technologies. While, for example, a separate fibre core with 20 Gbps capacity is required for every seven BSs in a fully centralized network corresponding to split A, more than 800 can be supported per fibre core when utilizing split D. Similarly, more heterogeneous technologies like wireless FH can only be utilized with higher-layer splits, thereby replacing the expensive fibre entirely on the last mile.

Starting from the physical layer technologies, future mobile networks have to aim to converge the BH and FH technologies and utilize unified methods across all network layers. They have to transport different types of traffic, meet a set of greatly different QoS parameters and be flexibly reconfigurable while utilizing different physical technologies. While the implementation of such a network will be challenging in the beginning, it offers the chance to reduce the complexity of network management and operation substantially while additionally reducing deployment cost, culminating in a virtualization of the transport network. In fact, a fully virtualized RAN should not only aim to virtualize a single part of the network like baseband processing, but all elements including base stations, fronthaul, baseband processors

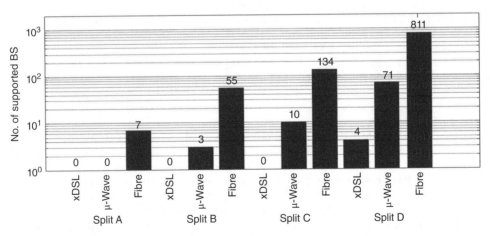

Figure 4.9 Number of supported BSs for different aggregation technologies: xDSL at 200 Mbps; μ-wave at 2 Gbps; fibre at 20 Gbps. Reproduced from [10] with permission from the IEEE

and backhaul. Numerous tools such as a flexible functional split, a converged BH/FH network, SDN, NFV and cloudlets will have to come together to achieve the flexibility that is required to improve performance, cost efficiency and adaptability of future mobile networks.

Acknowledgement

 The research leading to these results has received funding from the European Union's Horizon 2020 research and innovation programme under grant agreement No 671551 and the European Union's Seventh Framework Programme (FP7/2007-2013) under grant agreement No 317941. The European Union and its agencies are not liable or otherwise responsible for the contents of this document; its content reflects the view of its authors only.

References

[1] Guan, H., Kolding, T. and Merz, P. (2010) *Discovery of cloud-RAN*.

[2] Irmer, R., Droste, H., Marsch, P., Grieger, M., Fettweis, G., Brueck, S., Mayer, H.-P., Thiel, L. and Jungnickel, V. (2011) Coordinated multipoint: Concepts, performance, and field trial results. *IEEE Communications Magazine*, **49**(2), 102–111.

[3] Sawahashi, M., Kishiyama, Y., Morimoto, A., Nishikawa, D. and Tanno, M. (2010) Coordinated multipoint transmission/reception techniques for LTE-advanced. *IEEE Wireless Communications*, **17**(3), 26–34.

[4] Caire, G. and Müller, R. (2001) The Optimal Received Power Distribution of IC-Based Iterative Multiuser Joint Decoders. *Proceedings of the 39th Annual Allerton Conference on Communication, Control and Computing*, Monticello, IL, October.

[5] Rost, P., Bernardos, C. J., De Domenico, A., Di Girolamo, M., Lalam, M., Maeder, A., Sabella, D. and Wübben, D. (2014) Cloud Technologies for Flexible 5G Radio Access Networks. *IEEE Communications Magazine*, **52**(5), 68–76.

[6] Armbrust, M., Fox, A., Griffith, R., Joseph, A. D., Katz, R., Konwinski, A., Lee, G., Patterson, D., Rabkin, A., Stoica, I. and Zaharia, M. (2010) A view of cloud computing. *Communications ACM*, **53**(4), 58.

[7] Guo, B., Cao, W., Tao, A. and Samardzija, D. (2013) LTE/LTE-A Signal Compression on the CPRI Interface. *Bell Labs Technical Journal*, **18**(2), 117–133.

[8] Suryaprakash, V., Rost, P. and Fettweis, G. (2015) Are Heterogeneous Cloud-Based Radio Access Networks Cost Effective? *IEEE Journal on Selected Areas in Communication*, **33**(10), 2239–2251.

[9] Wübben, D., Rost, P., Bartelt, J., Lalam, M., Savin, V., Gorgoglione, M., Dekorsy, M. and Fettweis, G. (2014) Benefits and Impact of Cloud Computing on 5G Signal Processing: Flexible centralization through cloud-RAN. *IEEE Signal Processing Magazine*, **31**(6), 35–44.

[10] Bartelt, J., Rost, P., Wübben, D., Lessmann, J., Melis, B. and Fettweis, G. (2015) Fronthaul and Backhaul Requirements of Flexibly Centralized Radio Access Networks. *IEEE Wireless Communications*, **22**(5), 105–111.

[11] Common Public Radio Interface, CPRI Specification V6.0, 2013. Available at: http://www.cpri.info/. Accessed: September 1, 2015.

[12] Nadiv, R. and Naveh, T. (2010) Wireless Backhaul Topologies: Analyzing Backhaul Topology Strategies. Ceragon white paper, August. Available at: https://www.ceragon.com/images/Reasource_Center/White_Papers/Ceragon_Wireless_Backhaul_Topologies_Tree_vs_%20Ring_White_Paper.pdf. Accessed: October 8, 2015.

[13] 3GPP (2015) 3GPP TS 36.213, 'Evolved Universal Terrestrial Radio Access (E-UTRA); Physical layer procedures (Release 12),' v12.5.0, April.

[14] Valenti, M. C. (1999) *Iterative Detection and Decoding for Wireless Communications*. PhD dissertation, Virginia Polytechnic Institute and State University, Blacksburg, VA.

[15] Paul, H., Shin, B.-S., Wübben, D. and Dekorsy, A. (2013) In-network-processing for small cell cooperation in dense networks. *Proceedings of the IEEE 78th Vehicular Technology Conference*, Las Vegas, NV, September.

[16] Fritzsche, R., Rost, P. and Fettweis, G. (2015) Robust Rate Adaptation and Proportional Fair Scheduling With Imperfect CSI. *IEEE Transactions on Wireless Communication*, **14**(8), 4417–4427.

[17] De Domenico, A., Savin, V. and Ktenas, D. (2013) A backhaul-aware cell selection algorithm for heterogeneous cellular networks. *IEEE 24th International Symposium on Personal Indoor and Mobile Radio Communications*, London, September.

[18] Larsson, E., Edfors, O., Tufvesson, F. and Marzetta, T. (2014) Massive MIMO for next generation wireless systems. *IEEE Communications Magazine*, **52**(2), 186–195.

[19] Dötsch, U., Doll, M., Mayer, H. P., Schaich, F., Segel, J. and Sehier, P. (2013) Quantitative Analysis of Split Base Station Processing and Determination of Advantageous Architectures for LTE. *Bell Labs Technical Journal*, **18**(1), 105–128.

[20] Huang, J., Yuan, Y. *et al.* (2015) White Paper of Next Generation Fronthaul Interface. White paper, June. Available at: http://labs.chinamobile.com/cran/wp-content/uploads/White%20Paper%20of%20Next%20Generation%20Fronthaul%20Interface.PDF. Accessed: October 8, 2015.

[21] Khalili, S. and Simeone, O. (2015) Uplink HARQ for C-RAN via Low-Latency Local Feedback over MIMO Finite-Blocklength Links. arXiv preprint, arXiv:1508.06570.

[22] Rappaport, T. S. (2002) *Wireless Communications: Principles and Practice*, 2nd edition. New Jersey: Prentice Hall.

[23] d'Halluin, Y., Forsyth, P. A. and Vetzal. K. R. (2007) Wireless Network Capacity Management: A Real Options Approach. *European Journal of Operations Research*, **176**(1), 584–609.

[24] Rice, J. (2006) *Mathematical Statistics and Data Analysis*, 2nd edition, Belmont: Duxbury Press, pp. 181–187.

[25] Zervas, G., Triay, J., Amaya, N., Qin, Y., Cervelló-Pastor, C. and Simeonidou, D. (2011) Time Shared Optical Network (TSON): a novel metro architecture for flexible multi-granular services. *Optics Express*, **19**(26), B509–B514.

[26] ITU-T Recommendation G.9701, 'Fast access to subscriber terminals (G.fast) – Physical layer specification,' 2014.

[27] NFSO-ICT-317941 iJOIN, 'D5.2 – Final Definition of Requirements and Scenarios,' November 2014. Available at: http://www.ict-ijoin.eu/wp-content/uploads/2012/10/D5.2.pdf. Accessed: September 1, 2015.

[28] IEEE (2014) 'Standard for Radio over Ethernet Encapsulations and Mappings.' IEEE Standard P1904.3, 2014. Available at: http://standards.ieee.org/develop/project/1904.3.html. Accessed: September 1, 2015.

[29] IEEE (2013) 'Standard for a Precision Clock Synchronization Protocol for Networked Measurement and Control Systems.' IEEE Standard P1588-2008, 2013. Available at: https://standards.ieee.org/develop/project/1588.html. Accessed: September 1, 2015.

[30] IEEE (2014) 'Standard for Ethernet.' IEEE Standard P802.3-2012, 2014. Available at: http://standards.ieee.org/findstds/standard/802.3-2012.html. Accessed: September 1, 2015.

[31] IEEE(2014) 'Standard for Ethernet Amendment: Media Access Control Parameters, Physical Layers and Management Parameters for 400 Gb/s Operation.' IEEE Standard P802.3bs, 2014. Available at: http://www.ieee802.org/3/bs/index.html. Accessed: September 1, 2015.

[32] Fettweis, G. (2014) The Tactile Internet: Applications and Challenges. *IEEE Vehicular Technology Magazine*, **9**(1), 64–70.

[33] Open Networking Foundation (2012) Software-Defined Networking: The New Norm for Networks. White Paper, April. Available at: https://www.opennetworking.org/images/stories/downloads/white-papers/wp-sdn-newnorm.pdf. Accessed: September 1, 2015.

[34] ETSI Industry Specification Group for Network Functions Virtualisation (n.d.). Available at: http://www.etsi.org/technologies-clusters/technologies/nfv. Accessed: September 1, 2015.

[35] Mell, P. and Grance, T. (2011) *The NIST Definition of Cloud Computing*. US National Institute of Science and Technology. Available at: http://csrc.nist.gov/publications/nistpubs/800-145/SP800-145.pdf. Accessed: September 1, 2015.

[36] Li, L., Liu, J., Xion, K. and Butovitsch, P. (2012) Field test of uplink CoMP joint processing with C-RAN testbed. *7th International ICST Conference on Communication and Networking China*, August.

[37] Paul, H., Wübben, D. and Rost, P. (2015) Implementation and Analysis of Forward Error Correction Decoding for Cloud-RAN Systems. *2nd International Workshop on Cloud-Process. Heterogeneous Mobile Communication Networks*, London, UK, June.

[38] Rost, P., Talarico, S. and Valenti, M. C. (2015) The Complexity-Rate Tradeoff of Centralized Radio Access Networks. arXiv preprint, arXiv:1503.08585.

[39] Hewlett-Packard Company (2014) Efficient deployment of virtual network functions on HP ProLiant m800. Technical white paper. Available at: http://h20195.www2.hp.com/V2/getpdf.aspx/4AA5-5395ENW.pdf. Accessed: September 1, 2015.

[40] Huang, C. I. J., Duan, R., Cui, C., Jiang, J. X. and Li, L. (2014) Recent Progress on C-RAN Centralization and Cloudification. *IEEE Access*, **2**, 1030–1039.

[41] Satyanarayanan, M., Bahl, P., Caceres, R. and Davies, N. (2009) The case for VM-based cloudlets in mobile computing. *IEEE Pervasive Computing*, **8**(4), 14–23.

5

Analysis and Optimization for Heterogeneous Backhaul Technologies

Gongzheng Zhang,[1] Tony Q. S. Quek,[2] Marios Kountouris,[3] Aiping Huang[1] and Hangguan Shan[1]

[1]*College of Information Science and Electronic Engineering, Zhejiang University, China*

[2]*Information Systems Technology and Design Pillar, Singapore University of Technology and Design, Singapore*

[3]*Mathematical and Algorithmic Sciences Laboratory, France Research Centre, Huawei Technologies, France*

5.1   Introduction

Densifying the cellular network via deploying ultra-dense small-cell base stations (BSs) is a promising way to meet the tremendous demand for cellular data as we approach next generation cellular networks [1, 2]. To carry the traffic from BSs to the core network and vice versa, the backhaul network needs to be enhanced proportionally. Meanwhile, low delay on the radio access and backhaul links becomes essential to deliver a wide range of services and applications in future cellular networks, for example, VoIP and online gaming with an acceptable quality of service (QoS) [3]. As a result, backhaul has become the next big challenge for providing reliable and timely connectivity between BSs and the core network [4, 5], especially for delay-sensitive services or network functionalities.

Backhauling/Fronthauling for Future Wireless Systems, First Edition.
Edited by Kazi Mohammed Saidul Huq and Jonathan Rodriguez.
© 2017 John Wiley & Sons, Ltd. Published 2017 by John Wiley & Sons, Ltd.

Unlike traditional macro-cell BSs, which are usually directly connected to the operator's core network through fibre with very low latency or microwave links with high reliability, small-cell BSs are not always in easy-to-reach locations, for example near street level or lampposts rather than rooftops, which makes conventional fibre or microwave links impractical or cost-inefficient. Many wired and wireless technologies have been proposed as backhaul solutions for small cells [6–10]. Wired backhaul technologies have the advantages of high reliability, high data rates and high availability. However, they may sustain long and variable delays in the backbone routes or switches due to multiple hops, especially for xDSL, which can only reach up to 200–400 metres per single hop [11]. Wireless backhaul can be deployed more easily and at lower cost. Sub-6 GHz wireless backhaul has the advantage of non-line-of-sight (NLOS) transmission, while the presence of interference due to the co-existence issue makes the wireless links unreliable and introduces unpredictable delay. Millimetre-wave (mmWave) technologies of 60 GHz and 70–80 GHz are another potential backhaul solution, as they offer high capacity and reliability based on line-of-sight (LOS) links. Due to the small carrier wavelength and possibility of directional beamforming, the mmWave links can indeed be modelled as pseudo-wired without interference, which makes them extremely suitable for dense small-cell networks [12]. However, multi-hop implementation is needed in the absence of LOS, which causes additional delay. Due to these diverse characteristics, heterogeneous backhaul deployment will be a potential solution. It is essential to model and compare the performance of these different types of backhaul technologies to provide guidelines for such a system design.

Besides their capabilities, the cost of these backhaul technologies behaves differently in terms of both deployment and operation, which is another aspect that may limit the deployment of small-cell networks. Specifically, the deployment of a wired link (e.g. fibre or cable) is much more expensive than a wireless link, while the operational cost (e.g. the power consumption) of a wired link is much lower than a wireless link. As a result, it is quite a challenging task to identify the most appropriate and efficient solution for backhaul infrastructure, especially for dense small-cell networks. Furthermore, optimizing the configurations so as to minimize the cost is critical for the operators.

Furthermore, in a two-tier heterogeneous cellular network where a small-cell network is overlaid with a macro-cell network, BS association is another challenging problem. From a signal quality perspective, users will prefer to associate with macro-cell BSs due to their larger transmit power. By contrast, from a load point of view, users will prefer to associate with small-cell BSs, which are usually underutilized. Furthermore, imperfections in the backhaul link may result in packet delay increases if users associate with small-cell BSs. Therefore, the BS association policy should take the signal quality, load and backhaul into consideration to optimize the overall network performance.

As a timely and practically relevant topic, backhaul has been drawing much attention in recent years from many angles, including transmission, cost and system design. Various backhaul technologies are introduced in [4–10]. Modelling the delay performance of wired networks through measurements has a long history, and the results for routers and switches can be found in [13–17]. The transmission characteristics of mmWave and sub-6 GHz form another hot topic, and initial results for mmWave can be found in [12, 18–21]. There is some system design work taking backhaul into consideration, where the backhaul is considered as a capacity constraint [22, 23]. Some other work tries to estimate the backhaul cost via listing all the components of the network and their prices, which is practical but lacks theoretical analysis [24–27]. Cost models for some specific technologies are presented in [28–30]. The BS association problem is studied for heterogeneous cellular networks in [31–33]. Different joint spectrum-partitioning and bias-based BS association algorithms are proposed to maximize the user rate or rate coverage in [34–36]. However, backhaul is not considered in that work, even though it has a significant effect, as shown in our work, and can change the whole picture. As mentioned previously, a general backhaul model is needed, especially for studying the delay performance and designing the backhaul network to minimize cost.

The main contributions of this work are twofold. First, we propose a backhaul model for four promising backhaul technologies as a means to investigate the effect of different backhaul technologies on the network performance. Specifically:

- The packet delay in backhaul links is modelled for four technologies, including fibre, xDSL, mmWave and sub-6 GHz, each with distinct transmission characteristics.
- The mean packet delay and delay-limited success probability are analysed and compared for the above-mentioned technologies. Our results show that fibre is always the best choice, sub-6 GHz and xDSL are suitable for links with modest link length, while mmWave is a very competitive candidate for short links in terms of delay performance.
- A tractable model for quantifying the backhaul cost is presented and the mean backhaul cost per small-cell BS is analysed. A key result is that there exists an optimal gateway density which minimizes the mean cost, and the optimal operating point depends on the ratio of per gateway cost to per unit length link cost.

The second contribution of our work is to propose a backhaul-aware BS association policy for two-tier cellular networks. The aim is to minimize the mean network packet delay coming from both the backhaul and radio access links. Some remarkable conclusions are obtained from numerical results. Specifically:

- Backhaul delay may dominate the mean network packet delay from gateways to users when the backhaul network does not scale proportionally to the increasing number of small cells.
- The proposed backhaul-aware BS association policy outperforms conventional association policies with or without biasing, in terms of mean network packet delay.

- BS association without biasing may even outperform biased ones without considering backhaul, which implies that users may be misled into associating with small-cell BSs, thus deteriorating the system performance.

The remainder of this chapter is organized as follows. Section 5.2 presents the network model, the packet delay model and the cost model. The mean packet delay and delay-limited success probability of backhaul links are analysed in Section 5.3. The backhaul cost is analysed in Section 5.4. In Section 5.5, the mean network packet delay is analysed, based on which the backhaul-aware BS association is proposed and evaluated for a two-tier cellular network. Section 5.6 summarizes the chapter.

5.2   Backhaul Model

5.2.1   Network Model

We consider a two-tier cellular network consisting of radio access and backhaul networks, with gateways, hubs, macro-cell and small-cell BSs and users as components, as illustrated in Figure 5.1. The macro-cell BSs are always co-located with gateways, while the small-cell BSs connect to the gateways via various backhaul technologies. We denote the macro-cell network and small-cell network as tiers one and two, respectively. The locations of the gateways, macro-cell BSs, small-cell BSs and users are modelled as independent homogeneous Poisson point processes (PPPs), Φ_g, $\Phi_{b,1}$, $\Phi_{b,2}$ and Φ_u of density λ_g, $\lambda_{b,1}$, $\lambda_{b,2}$ and λ_u, respectively. Without loss of generality,

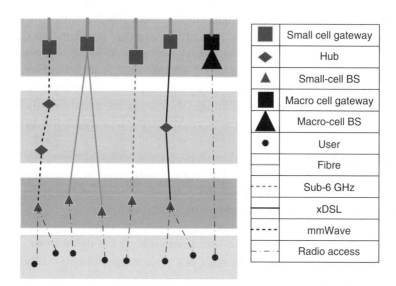

Figure 5.1   Heterogeneous network model embracing various backhaul technologies. Reproduced from [43] with permission from the IEEE

the links from gateways to the core network are assumed to be a common infrastructure for all four backhaul technologies and thus are neglected in the following.

The radio access network connects users with (macro-cell or small-cell) BSs through wireless links, which usually have only one hop. The backhaul network is composed of links connecting small-cell BSs with gateways, which may be multi-hop links depending on the type of technologies used. The backhaul links using different technologies may differ in the number of hops due to different transmission ranges per single hop. We denote the transmission range of one hop in the backhaul link by r, which is the distance to guarantee a certain minimum capacity. For instance, the transmission ranges of popular backhaul technologies can generally be ordered as $r_{\text{fibre}} > r_{\text{sub-6 GHz}} > r_{\text{xDSL}} > r_{\text{mmWave}}$ [7]. We denote the number of hops along the link by n, which is determined by the link length d as $n = \lceil d/r \rceil$. In the following, we consider that the small-cell BS is associated with the nearest gateway. In this case, the length of a backhaul link follows a Rayleigh distribution with probability density function (pdf) given by [37]:

$$f_D(d) = 2\pi\lambda_g d\exp\left(-\pi\lambda_g d^2\right).$$

(5.1)

Therefore, the average length of the backhaul link can be obtained as $1/\sqrt{2\lambda_g}$, and the average number of hops in a backhaul link $\bar{n}$ (≥ 1) can be estimated as [37]:

$$\bar{n} = \frac{1}{r\sqrt{2\lambda_g}}.$$

(5.2)

This is an optimistic estimation since the link length is defined as the physical distance between the BS and the gateway.

5.2.2   Delay Model

In this work, we focus on the packet delay from the gateway to the user, that is, the downlink scenario. The packet delay has a significant impact on the queueing and end-to-end delays, and it consists of the packet transmission and propagation delays along the link as well as of the processing delay at each node.

- For wired backhaul, the packet delay mainly comes from the processing time in the gateway and hubs, which means that the transmission and propagation delays can be ignored due to the relatively large capacity and highly reliable transmission of the wired backhaul.
- For wireless backhaul, the packet delay along the link mainly comes from the transmission time in each hop in case of retransmissions, because the decode-and-forward procedure is typically applied in each hop.

5.2.2.1   Delay in Wired Links

In wired scenarios, the exponential distribution has been used to model the delay in routers and switches [28, 38]. However, many measurements indicate that the distribution of the router or switch delay behaves as gamma distributed or long-/heavy-tailed. Nevertheless, there are various types of distributions that fit for different specific types of router, including Gaussian, gamma, Weibull and Pareto distributions [13–17]. In the following, a general but tractable distribution is needed to model the processing delay of the nodes in wired links. As a result, we assume that the processing delay in each hop follows a gamma distribution with parameters depending on the load, which includes the exponential distribution as a special case.

Given the number of hops of the link, n, a packet in the backhaul has to traverse through a gateway and $n-1$ hubs. We denote the processing delay in the gateway and the jth ($j=1,\ldots,n-1$) hub as T_g and $T_{h,j}$, respectively. Then, the total backhaul delay for the BS is given by:

$$T_{\mathrm{bh,wd}} = T_g + \sum_{j=1}^{n-1} T_{h,j} \tag{5.3}$$

where the subscript wd is used to denote wired connections. The processing delay in the gateway depends on the size of the packet and the number of small-cell BSs associated with the gateway. We model T_g using the gamma distribution with parameters depending on the mean number of small-cell BSs associated with a gateway as follows:

$$T_g \sim \mathrm{Gamma}\left(\left(1+1.28\frac{\lambda_{b,2}}{\lambda_g}\right)\kappa_1, a+b\mu\right) \tag{5.4}$$

where $1+1.28\dfrac{\lambda_{b,2}}{\lambda_g}$ denotes the mean number of small-cell BSs in the gateway covering the chosen BS [32]. Specifically, the first and second terms represent the effect of the number of connecting nodes and packet size on delay, respectively. In addition, a, μ and κ_1 are constants that reflect the processing capability of the nodes, and b is the packet size. The exact values of these parameters can be obtained through fitting using real measurements, which goes beyond our scope. Since a chain topology is adopted, there is one ingress and one egress in each hub. The delay in each hub is assumed independent and follows a gamma distribution with an identical parameter κ_2 as follows:

$$T_{h,j} \sim \mathrm{Gamma}(\kappa_2, a+b\mu), \; j=1,2,\ldots,n-1. \tag{5.5}$$

Therefore, given the number of hops, n, the total backhaul delay in the wired backhaul link still follows the gamma distribution as follows:

$$T_{\mathrm{bh,wd}} \sim \mathrm{Gamma}\left(\left(1+1.28\frac{\lambda_{b,2}}{\lambda_g}\right)\kappa_1 + (n-1)\kappa_2, a+b\mu\right). \tag{5.6}$$

5.2.2.2 Sub-6 GHz Wireless Links

For the sub-6 GHz wireless backhaul case, we consider that a dedicated spectrum with bandwidth of $W_{\text{sub-6 GHz}}$ is allocated to backhaul links. Setting the receiver at the origin, the received signal power from the transmitter located at x is $P_Y h_x |x|^{-\alpha}$, where P_Y and α are the transmit power and path-loss exponent, respectively, and the subscript Y represents the type of node, that is, gateway, hub or BS. Here, gateways and hubs are assumed to have the same transmit power to reach the same transmission range. In the following, h_x is the small-scale fading (channel gain) and Rayleigh fading with unit mean. Furthermore, an interference-limited scenario is considered, that is, the effect of background noise is neglected. If the intended transmitter is located at x_o, the received signal-to-interference ratio (SIR) can be expressed as:

$$\text{SIR} = \frac{P_Y h_{x_o} |x_o|^{-\alpha}}{\sum_{x \in \Phi \backslash x_o} P_Y h_x |x|^{-\alpha}} \tag{5.7}$$

where $\Phi \backslash x_o$ is the set of interfering nodes that use the same spectrum. Specially, the interfering nodes for a sub-6 GHz gateway are the other gateways using sub-6 GHz technologies.

The wireless transmission (including sub-6 GHz and mmWave) is time slotted and one packet is transmitted in each time slot. The transmission in one hop succeeds if the received SIR is above a threshold, θ. Otherwise, the transmission fails and retransmission is required. Obviously, the transmission success probability in a single transmission attempt depends on the link length r and interfering nodes $\Phi \backslash x_o$. Considering Gaussian codebooks, the amount of bits that can be transmitted in a single successful transmission is given by:

$$b = \tau_{\text{bh,s6}} W_{\text{sub-6 GHz}} \log_2(1+\theta) \tag{5.8}$$

where $\tau_{\text{bh,s6}}$ is the time slot length with the subscript s6 representing sub-6 GHz. The packet delay in sub-6 GHz backhaul links is the time that is needed to successfully transmit a packet from a gateway to a small-cell BS via a sub-6 GHz link.

5.2.2.3 mmWave Links

The propagation characteristic of a mmWave link is quite different from a sub-6 GHz link. Due to the relatively high frequency in the mmWave band, LOS is needed to establish a link, where one-hop transmission can reach about 100–200 m [18, 20, 39]. However, the high frequency also leads to the possibility of directional antennas, which results in a pseudo-wired noise-limited rather than an interference-limited network [21]. Applying the fitted model in [18], the path loss in an LOS link with distance r is given by:

$$L(\text{dB}) = 70 + 20 \log_{10}(r+\xi), \xi \sim N(0,\sigma^2) \tag{5.9}$$

where ξ is the shadow fading coefficient and σ is the standard deviation of shadow fading in dB [20]. Denote the transmit power plus the antenna gains by P_{tx}(dB) and the noise power density by N_0. The transmission in one hop succeeds if the received signal-to-noise ratio (SNR) is larger than θ, that is:

$$P_{tx}(dB) - L(dB) - N_0 W_{mmWave}(dB) \geq \theta(dB) \tag{5.10}$$

where W_{mmWave} is the bandwidth and retransmission is required when failure occurs.

The packet delay in mmWave backhaul links is the time that is needed to successfully transmit a packet from a gateway to a small-cell BS via a mmWave link. If the distance between the small-cell BS and the gateway is larger than the one-hop maximum transmission range of a single hop, the mmWave backhaul needs to be deployed with multiple hops. For multi-hop mmWave backhaul links, a decode-and-forward protocol is adopted and the total backhaul delay is simply the summation of delay in all the hops.

5.2.3 Cost Model

The total cost to run the network can generally be broken down into two categories, namely, the capital expenditure (CAPEX) and operational expenditure (OPEX). The former is a one-off insertion cost, mainly including the equipment cost of BSs, gateways, hubs and links, the installation cost of these infrastructures, etc. The latter mainly consists of the annual power consumption bill to operate the network and other maintenance costs, which can be estimated as a percentage of the CAPEX [24]. Therefore, we only focus on the CAPEX in this work, and the other aspects can be obtained similarly. Furthermore, we only consider the backhaul cost including the gateway and backhaul link cost, because the small-cell BS density is determined by the subscriber requirement and the dimensioning method [40].

The gateway cost is determined by the number of deployed gateways and the number of small-cell BSs. The cost to deploy one gateway is composed of the installation cost including the site rental and the capacity cost including the configured transceivers, one for each small-cell BS. We denote the cost of deploying a gateway of type z by $C_{gw,z}$, which is then given by:

$$C_{gw,z} = U_{z,0} + U_{z,1} N_{BS} \tag{5.11}$$

where N_{BS} is the number of small-cell BSs connecting to the gateway and $U_{z,0}$ and $U_{z,1}$ are technology-specific constants. Here, gw represents gateway, and $z = f, x, s, m$ represents fibre, xDSL, sub-6 and mmWave, respectively.

The link cost is a function of the length of the link for each technology. We denote the cost of deploying a backhaul link of the zth type by $C_{lk,z}$, which is given by:

$$C_{lk,z} = V_{z,0} d^{\beta_{z,0}} + V_{z,1} d^{\beta_{z,1}} \tag{5.12}$$

where lk represents link, and $V_{z,0}$ and $V_{z,1}$ are technology-specific constants. Here, $V_{z,0}d^{\beta_{z,0}}$ represents the infrastructure cost of deploying the link, which is usually a linear function of the link length for the wired backhaul ($\beta_{z,0} = 1$) while it can be ignored for the wireless backhaul ($V_{z,0} = 1$). $V_{z,1}d^{\beta_{z,1}}$ represents the capacity cost, where the index $\beta_{z,1}$ is also usually 1 for the wired backhaul but may be 2–6 for the wireless backhaul.

5.3 Backhaul Packet Delay Analysis

The network packet delay is defined as the time needed for a packet to be successfully received by a user, which is the summation of the backhaul delay and radio access delay. In this section, we present the main results on the mean packet delay and the delay-limited success probability in different backhaul links.

5.3.1 Mean Backhaul Packet Delay

5.3.1.1 Wired Backhaul

Proposition 1: *The conditional mean packet delay (conditioned on the number of hops, n) in wired backhaul links is given by:*

$$T_{\text{bh,wd}} = \left(\left(1 + 1.28 \frac{\lambda_{b,2}}{\lambda_g} \right) \kappa_1 + (n-1)\kappa_2 \right)(a + b\mu). \tag{5.13}$$

The mean packet delay in wired backhaul is given by:

$$\overline{T_{\text{bh,wd}}} \approx \left(\left(1 + 1.28 \frac{\lambda_{b,2}}{\lambda_g} \right) \kappa_1 + \left(\frac{1}{r\sqrt{2\lambda_g}} - 1 \right) \kappa_2 \right)(a + b\mu). \tag{5.14}$$

Proof: Equation (5.13) can be directly obtained from the expectation of the gamma distribution in Equation (5.6). Applying the approximation of the mean number of hops of a backhaul link in Equation (5.2), we obtain Equation (5.14).

5.3.1.2 Sub-6 GHz Backhaul

Proposition 2: *The conditional mean packet delay (conditioned on the distance between the BS and its nearest gateway, d) in sub-6 GHz backhaul links is given by:*

$$T_{\text{bh,s6}} = \tau_{\text{bh,s6}} \left(1 + 1.28 \frac{\lambda_{b,2}}{\lambda_g} \right) \exp(\pi \lambda_g \rho(\alpha,\theta)d^2) \tag{5.15}$$

where $\rho(\alpha,\theta) = \theta^\delta \int_{\theta^{-\delta}}^{\infty} \frac{1}{1+u^{1/\delta}} \, du$ and $\delta = 2/\alpha$. *The mean packet delay in sub-6 GHz backhaul links is given by:*

$$\overline{T_{bh,s6}} = \tau_{bh,s6} \left(1+1.28\frac{\lambda_{b,2}}{\lambda_g}\right)\left(1+\rho(\alpha,\theta)\right). \tag{5.16}$$

Proof: Given the distance between the BS and its nearest gateway, d, the interfering nodes are outside of the ball centred at the BS with radius d. The transmission success probability in a single transmission attempt is given by [37]:

$$P_{ss,s6} = \exp(-\pi\lambda_g \rho(\alpha,\theta) d^2). \tag{5.17}$$

The mean number of transmissions to successfully deliver the packet is, in turn, $1/p_{ss,s6}$. Furthermore, the mean number of small-cell BSs in a gateway covering the chosen BS is $1+1.28\frac{\lambda_{b,2}}{\lambda_g}$, so the probability that the gateway transmits to this BS in each time slot is given by:

$$p_{sl} = \frac{\lambda_g}{\lambda_g + 1.28\lambda_{b,2}}. \tag{5.18}$$

Therefore, the mean number of time slots to successfully deliver the packet is $1/(p_{sl}p_{ss,s6})$, which, multiplied by the time slot length gives Equation (5.15). Taking expectation with respect to the random distance d with the pdf given by Equation (5.1) leads to Equation (5.16).

5.3.1.3   mmWave Backhaul

Proposition 3: *Given that each hop has a constant distance, r, the conditional mean packet delay (conditioned on the number of hops, n) in mmWave backhaul links is given by:*

$$T_{bh,mm} = \left(1+1.28\frac{\lambda_{b,2}}{\lambda_g}\right)\frac{2n\tau_{bh,mm}}{1+\text{erf}\left(\frac{\theta'(r)}{\sqrt{2}\sigma}\right)} \tag{5.19}$$

where $\text{erf}(\cdot)$ *is the error function and* $\theta'(r)(dB) = P_{tx}(dB) - \theta(dB) - N_0 W_{mmWave}$ $(dB) - 70 - 20\log_{10} r$. *The mean packet delay in mmWave backhaul links is approximated as:*

$$\overline{T_{bh,mm}} \approx \left(1+1.28\frac{\lambda_{b,2}}{\lambda_g}\right)\frac{2\tau_{bh,mm}}{r\sqrt{2\lambda_g}\left(1+\text{erf}\left(\frac{\theta'(r)}{\sqrt{2}\sigma}\right)\right)}. \tag{5.20}$$

Proof: From Equation (5.10), the probability that a transmission of a single hop in a single time slot succeeds is given by:

$$P_{ss,mm} = Pr\left(L \leq P_{tx}(dB) - \theta(dB) - N_0 W_{mmWave}(dB)\right). \tag{5.21}$$

Given the hop distance, r, the transmission success probability can be obtained from Equation (5.9) as:

$$\begin{aligned} P_{ss,mm} &= Pr\left(\xi \leq P_{tx}(dB) - \theta(dB) - N_0 W_{mmWave}(dB) - 70 - 20\log_{10} r\right) \\ &= \frac{1}{2}\left(1 + erf\left(\frac{\theta'(r)}{\sqrt{2}\sigma}\right)\right) \end{aligned} \tag{5.22}$$

where $\theta'(r)(dB) = P_{tx}(dB) - \theta(dB) - N_0 W_{mmWave}(dB) - 70 - 20\log_{10} r$. Therefore, the mean packet delay in each hop is given by $1/(p_{sl}P_{ss,mm})$. Finally, multiplying by the number of hops and time slot length gives the conditional mean packet delay in the mmWave backhaul link as Equation (5.19). Applying the approximation of the mean number of hops of a backhaul link in Equation (5.2) leads to Equation (5.20).

5.3.2 Delay-limited Success Probability

To evaluate the capability of backhaul infrastructure for supporting traffic with delay requirements, we define a key performance metric for delay-sensitive services, coined as the *delay-limited success probability* and denoted by dp. The delay-limited success probability is the probability that a packet can be successfully delivered before a certain delay deadline. The deadline, denoted by t, is divided into the transmission deadlines for the backhaul part t_{bh} and the radio access part t_{ran} [5], that is:

$$t = t_{bh} + t_{ran}. \tag{5.23}$$

For time-slotted wireless transmission including wireless backhaul and radio access links, the transmission deadline can be translated into a scheduling and transmission constraint. The maximum number of time slots in which a packet in a wireless backhaul link can be scheduled and transmitted is given by:

$$k_{bh} = \lceil t_{bh}/\tau_{bh}\rceil. \tag{5.24}$$

Here, τ_{bh} is the common nomenclature for time slot lengths of various backhaul technologies.

Now, the delay-limited success probability of the backhaul link is defined as:

$$dp = \begin{cases} Pr\{T_{bh} \leq t_{bh}\}, & \text{wired backhaul} \\ Pr\{K_{bh} \leq k_{bh}\}, & \text{wireless backhaul} \end{cases} \tag{5.25}$$

where K_{bh} is the total number of transmissions (including retransmissions) for one packet along the link.

5.3.2.1 Wired Backhaul

Proposition 4: *Given the number of hops, n, the delay-limited success probability in the wired backhaul is given by:*

$$dp_{bh,wd} = \frac{\gamma\left(\left(1+1.28\frac{\lambda_{b,2}}{\lambda_g}\right)\kappa_1 + (n-1)\kappa_2, \frac{t_{bh}}{a+b\mu}\right)}{\Gamma\left(\left(1+1.28\frac{\lambda_{b,2}}{\lambda_g}\right)\kappa_1 + (n-1)\kappa_2\right)} \quad (5.26)$$

where $\gamma(s,x) = \int_0^x y^{s-1}e^{-y}ds$ is the incomplete gamma function and $\Gamma(s) = \int_0^\infty y^{s-1}e^{-y}ds$ is Euler's gamma function.

Proof: The probability that a packet is successfully delivered to a BS is equal to the probability that the packet delay along the link does not exceed the deadline. As the wired backhaul delay follows the gamma distribution, the probability can be directly obtained as:

$$\Pr\left(T_{bh,wd} < t_{bh}\right) = \frac{\gamma\left(\left(1+1.28\frac{\lambda_{b,2}}{\lambda_g}\right)\kappa_1 + (n-1)\kappa_2, \frac{t_{bh}}{a+b\mu}\right)}{\Gamma\left(\left(1+1.28\frac{\lambda_{b,2}}{\lambda_g}\right)\kappa_1 + (n-1)\kappa_2\right)} \quad (5.27)$$

which gives the result.

5.3.2.2 Sub-6 GHz Backhaul

Proposition 5: *Given the distance between the BS and the nearest gateway, d, the delay-limited success probability in sub-6 GHz backhaul is given by:*

$$dp_{bh,s6} = \sum_{j=1}^{k_{bh}} \binom{k_{bh}}{j}(-1)^{j+1}\left(\frac{\lambda_g}{\lambda_g + 1.28\lambda_{b,2}}\right)^j \exp\left(-j\pi\lambda_g\rho(\alpha,\theta)d^2\right). \quad (5.28)$$

Proof: Since the probability that the gateway transmits to the BS and the transmission succeeds in a single time slot is $p_{sl}P_{ss,s6}$, the delay-limited success probability is the probability that, in at least one of the K_{bh} time slots, the BS is scheduled and the transmission succeeds, that is:

$$dp_{bh,s6} = 1 - \left(1 - p_{sl}P_{ss,s6}\right)^{k_{bh}}$$
$$= \sum_{j=1}^{k_{bh}} \binom{k_{bh}}{j}(-1)^{j+1} p_{sl}^{j}P_{ss,s6}^{j}$$

(5.29)

where the second equality follows from the binomial expansion. Substituting Equation (5.17) into Equation (5.29) gives the result.

5.3.2.3 mmWave Backhaul

Proposition 6: *Given the number of hops, n, the delay-limited success probability in mmWave backhaul is given by:*

$$dp_{bh,mm} = \sum_{j=n}^{k_{bh}}\sum_{m=j}^{k_{bh}} \binom{k_{bh}}{j}\binom{k_{bh}-j}{m-j}(-1)^{m-j}\left(\frac{\lambda_g}{2(\lambda_g+1.28\lambda_{b,2})}\right)^{m}\left(1+\mathrm{erf}\left(\frac{\theta'(r)}{\sqrt{2}\sigma}\right)\right)^{m}.$$

(5.30)

Proof: Given the number of hops, n, the probability that the packet is successfully delivered to the BS before the delay deadline is equal to the probability that, in at least n out of the K_{bh} time slots, the BS is scheduled and the transmissions succeed, that is:

$$dp_{bh,mm} = \sum_{j=n}^{k_{bh}} \binom{k_{bh}}{j}\left(p_{sl}P_{ss,mm}\right)^{j}\left(1-p_{sl}P_{ss,mm}\right)^{k_{bh}-j}$$
$$= \sum_{j=n}^{k_{bh}}\sum_{m=j}^{k_{bh}} \binom{k_{bh}}{j}\binom{k_{bh}-j}{m-j}(-1)^{m-j} p_{sl}^{m}P_{ss,mm}^{m}$$

(5.31)

where the second equality follows from the binomial expansion. Substituting Equation (5.22) into Equation (5.31) gives the result.

5.3.3 Performance Evaluation

To evaluate the effects of the number of hops and packet size on the backhaul delay performance, we provide numerical results based on our analytical results derived in this section. The parameter settings are given in Table 5.1, where the time slot lengths

Table 5.1 Parameter settings. Reproduced from [43] with permission from the IEEE

Parameter	Value	Parameter	Value
λ_g	$10^{-7}/m^2$	θ	$0.1 \sim 50$
$\lambda_{b,2}$	$10^{-6}/m^2$	μ	0.01 μs/bit
r_{xDSL}	200 m	α	3.5
r_{mmWave}	100 m	σ	5
W_{mmWave}	200 MHz	κ_1	1
$W_{sub-6\,GHz}$	40 MHz	κ_2	10
$\tau_{bh,mm}$	5 μs	n	$1 \sim 20$
a	10 μs	$\tau_{bh,s6}$	25 μs
P_{tx}	30 dBm	t_{bh}	100 μs ~ 2 ms
N_0	−174 dBm/Hz		

are set to ensure the number of transmission bits in a packet given by Equation (5.8) is the same for different technologies. The parameter settings for wired backhaul are set to ensure the same transmission rate for the four backhaul technologies, so that we can focus on the effect of their characteristics on the delay performance. One hop is assumed for fibre and sub-6 GHz backhauls because the considered distance between the BS and the gateway is generally not large. The results for xDSL are obtained by replacing r with the one-hop transmission distance r_{xDSL} in Equation (5.14).

Figure 5.2 presents the mean backhaul packet delay under different transmission thresholds (accordingly, packet size) and distances (numbers of hops) for the considered four technologies. The following observations can be obtained from Figure 5.2(a):

- The mean backhaul delay increases with the transmission threshold, due to the increase of the packet size given by Equation (5.8), which increases the processing delay in wired backhaul or increases the mean number of retransmissions in wireless backhaul.
- The mean packet delay in wired backhaul increases approximately linearly with the transmission threshold, which represents the linear relationship with the packet size in our proposed model. However, for the wireless backhaul, the mean packet delay increases slightly in the low transmission threshold region but significantly in the high threshold region. This is because in the low transmission threshold region, the wireless backhaul transmits with a success probability approaching approximately 1, therefore the delay only depends on the time slot length and number of hops. On the other hand, in the high threshold region, the success probability decreases quickly and thus the number of retransmissions increases significantly.
- Fibre is always the best choice under the parameter settings, due to its longer reach range and higher transmission reliability. xDSL is better than wireless backhaul in the low and high transmission regions due to the short processing delay in xDSL

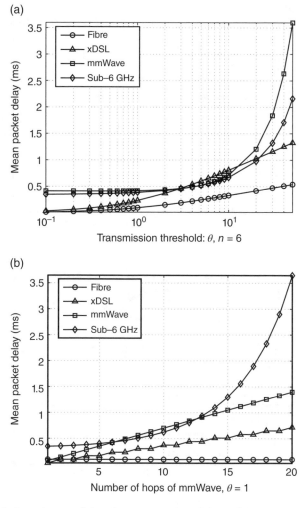

Figure 5.2   Variations in conditional mean packet delay of backhaul links with system parameters. Reproduced from [43] with permission from the IEEE

and highly unreliable transmission of wireless backhaul, respectively. Sub-6 GHz is more robust than mmWave due to their different characteristics, where the former needs to combat interference, as indicated in Equation (5.7) while the latter needs to combat random shadowing, as indicated in Equation (5.9).

Figure 5.2(b) provides the relationship between the mean packet delay and the distance. The distance is normalized by one hop length of mmWave to allow easy comparison between different backhaul technologies. The *x*-axis of Figure 5.2(b) has a range of [1~20], meaning that the distance is ranging from one to twenty hop

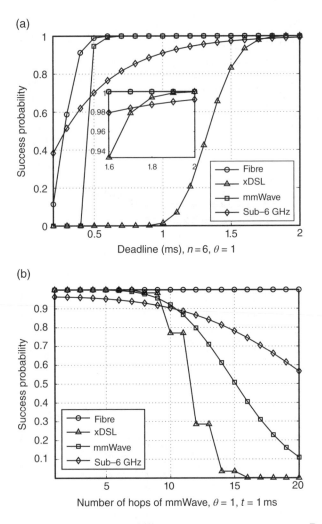

Figure 5.3 Delay-limited success probability versus system parameters. Reproduced from [43] with permission from the IEEE

lengths of mmWave. From Figure 5.2(b), we can see that the mean backhaul packet delay increases linearly with the number of hops for mmWave and xDSL but with different rates. On the other hand, even though sub-6 GHz has a longer transmission range, the mean packet delay performance is not always better due to its lack of transmission reliability, especially when the transmission distance is long. It can be observed from Figure 5.2(b) that when the distance between the BS and the gateway is quite short ($n = 1, 2$), mmWave's delay is a little higher than xDSL and fibre but lower than sub-6 GHz, hence mmWave seems to be a competitive candidate technology for such a scenario.

Figure 5.3 presents the delay-limited success probability of the four backhaul technologies. The success probabilities under different transmission deadlines are shown in Figure 5.3(a), which actually gives the cumulative density function of delay. The following observations can be obtained:

- Fibre is always the best choice as it has the highest delay-limited success probability. Sub-6 GHz is better than mmWave in the low-delay regime. This shows the advantage of direct transmission as compared with multi-hop transmission in terms of delay performance.
- Despite its low mean delay, as shown in Figure 5.2(b), xDSL is not an appropriate technology for backhaul when the transmission deadline is strict. Obviously, the delay-limited success probability of sub-6 GHz backhaul increases much slower as compared with other technologies, due to its lack of transmission reliability. mmWave outperforms xDSL because of the high processing delay dynamics of xDSL.

As can be observed from Figure 5.3(b), the success probability within the transmission deadline of xDSL is relatively poor when the number of hops is large. The delay-limited success probability of mmWave decreases more quickly with the number of hops as compared with sub-6 GHz technology. This, again, reveals the drawback of multi-hop transmission in terms of higher packet delay.

5.4 Backhaul Deployment Cost Analysis

In this section, we analyse the mean backhaul cost per small-cell BS. This framework is of wider interest, for example, it can apply to the cost estimation and optimization of cloud radio access networks by replacing the gateway and backhaul links with baseband processing units and fronthaul links, respectively.

Due to different capabilities of the different types of backhaul technologies and their deployment as well as operational costs, deploying fibre backhaul is expected to become more and more prevalent in the future, while xDSL technologies will be rarely used [6, 10]. However, due to the ultra-high density of small cells, sub-6 GHz will still be a supplement, especially for providing backhaul connectivity for small-cell BSs in hard-to-reach locations. Therefore, we model the locations of the BSs connecting to gateways with fibre, mmWave, xDSL and sub-6 GHz backhauls as independent thinning of $\Phi_{b,2}$, with probabilities p_f, p_m, p_x and p_s, respectively. As stated in Section 5.2.1, the small-cell BS is assumed to associate with the nearest gateway, no matter which type of backhaul it is.

Proposition 7: *The mean backhaul cost per small-cell BS is given by:*

$$\bar{C} = \sum_z p_z \left(U_{z,1} + U_{z,0} \frac{\lambda_g}{\lambda_{b,2}} + V_{z,0} \frac{\Gamma(\beta_{z,0}/2+1)}{\left(\pi\lambda_g\right)^{\beta_{z,0}/2}} + V_{z,0} \frac{\Gamma(\beta_{z,1}/2+1)}{\left(\pi\lambda_g\right)^{\beta_{z,1}/2}} \right) \qquad (5.32)$$

where the summation is taken with respect to the types of deployed backhaul technologies.

Proof: The mean backhaul cost per small-cell BS can be calculated as the mean backhaul cost per unit area divided by the small-cell BS density. The mean backhaul cost per unit area is the summation of the mean gateway cost per unit area and the mean link cost per unit area. Since the small-cell BS associates with the nearest gateway, the probability of the link deploying with the zth technology is p_z. The mean gateway cost per unit area is obtained as:

$$
\begin{aligned}
C_{gw} &= \frac{\lambda_g}{|\Phi_g|} \sum_{\Phi_g} (U_{z,0} + U_{z,1} N_{BS}) \\
&= \sum_z p_z \left(\lambda_g U_{z,0} + \lambda_{b,2} U_{z,1} \right).
\end{aligned}
\tag{5.33}
$$

The distribution of the link length is given by Equation (5.1). When the form of link cost is Vd^β, the mean cost is given by:

$$
\int_0^\infty Vd^\beta f_D(d) \, \mathrm{d}d = V \frac{\Gamma(\beta/2 + 1)}{\left(\pi \lambda_g \right)^{\beta/2}}.
\tag{5.34}
$$

Therefore, the mean link cost per unit area is given by:

$$
\begin{aligned}
C_{lk} &= \frac{\lambda_{b,2}}{|\Phi_{b,2}|} \sum_{\Phi_{b,2}} (V_{z,0} d^{\beta_{z,0}} + V_{z,1} d^{\beta_{z,1}}) \\
&= \sum_z p_z \lambda_{b,2} \left(\int_0^\infty V_{z,0} d^{\beta_{z,0}} f_D(d) \, \mathrm{d}d + \int_0^\infty V_{z,1} d^{\beta_{z,1}} f_D(d) \, \mathrm{d}d \right) \\
&= \sum_z p_z \lambda_{b,2} \left(V_{z,0} \frac{\Gamma(\beta_{z,0}/2 + 1)}{\left(\pi \lambda_g \right)^{\beta_{z,0}/2}} + V_{z,0} \frac{\Gamma(\beta_{z,1}/2 + 1)}{\left(\pi \lambda_g \right)^{\beta_{z,1}/2}} \right).
\end{aligned}
\tag{5.35}
$$

The mean backhaul cost per unit area is then $C = C_{gw} + C_{lk}$. Dividing C by the density $\lambda_{b,2}$ gives the mean backhaul cost per small-cell BS as Equation (5.32).

Note from Equation (5.32) that with increasing gateway density, the gateway cost increases linearly, while the link cost decreases. Therefore, there is a trade-off between the gateway cost and link cost, and there may be an optimal gateway density to minimize the mean backhaul cost per small-cell BS. This is further investigated below by explicitly considering some special cases.

Proposition 8: *Suppose a wired technology is solely adopted for the backhaul, for which the link cost increases linearly with the link length, that is, $\beta_0 = \beta_1 = 1$. The mean backhaul cost per small-cell BS is simplified as:*

$$
\overline{C}_{wd} = U_1 + U_0 \frac{\lambda_g}{\lambda_{b,2}} + V \frac{1}{2\sqrt{\lambda_g}}.
\tag{5.36}
$$

An optimal gateway density exists to minimize the mean cost, and is given by:

$$\lambda_g^* = \left(\frac{V}{4U_0}\right)^{2/3} \lambda_{b,2}^{2/3}.$$ (5.37)

Proposition 9: *Suppose a wireless technology is solely adopted for the backhaul, for which the infrastructure cost is ignored and the index of the capacity cost equals the path-loss exponent, that is, $\beta_1 = \alpha$. The mean backhaul cost per small-cell BS is simplified as:*

$$\overline{C_{wl}} = U_1 + U_0 \frac{\lambda_g}{\lambda_{b,2}} + V \frac{\Gamma(\alpha/2+1)}{\left(\pi\lambda_g\right)^{\alpha/2}}.$$ (5.38)

An optimal gateway density exists to minimize the mean cost, and is given by:

$$\lambda_g^* = \frac{V\alpha\Gamma\left(\alpha/2+1\right)^{\alpha/2}}{2U_0\pi^{\alpha/2}} \lambda_{b,2}^{\frac{2}{2+\alpha}}.$$ (5.39)

Figure 5.4 gives an illustration of mean backhaul cost per small-cell BS under different gateway densities for wired and wireless backhaul. The parameters are set as $V = 10^5$ and $\lambda_{b,2} = 10^{-5}/m^2$. The mean backhaul cost decreases quickly first and may increase gradually with gateway density. The quick decrease is due to the rapid decrease of the link length then the link cost when the gateway density increases. The slow increase is due to the increased gateway cost, which will finally exceed the decreased link cost. Obviously, the optimal gateway density decreases with the per-gateway cost. Furthermore, the optimal gateway density with wireless backhaul is much higher as compared with wired backhaul. This is mainly due to the higher link cost exponent of the wireless backhaul [$\beta_1 = 2$ (wireless) > 1 (wired)], which weakens the influence of the gateway cost.

5.5 Backhaul-aware BS Association Policy

In this section, we first give the mean network packet delay and then formulate and solve the optimal BS association problem. As pointed out in [31], cell range expansion based on bias is a suboptimal but quite efficient method to achieve load balancing from a network point of view, and can easily be implemented in the wireless standards. Therefore, we adopt this scheme in BS association.

5.5.1 Mean Network Packet Delay

5.5.1.1 Mean Radio Access Delay

For the radio access link, the delay also mainly comes from the transmission time.

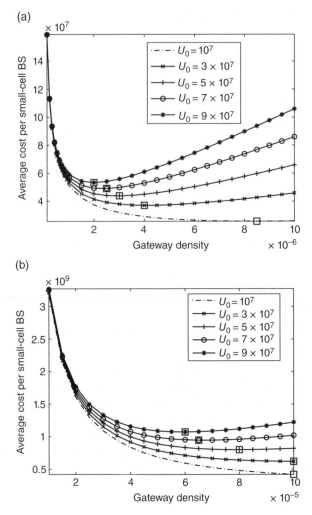

Figure 5.4 Variation in mean backhaul cost per small-cell BS with gateway density. Reproduced from [43] with permission from the IEEE

Consider a typical user located at the origin. The location of the nearest BS in the ith tier is denoted by $x_i^*, i = 1, 2$. Then, the user associates with the BS from which it receives the largest biased average power, that is, the tier is given by:

$$k = \arg \max_i B_i P_i \left| x_i^* \right|^{-\alpha} \qquad (5.40)$$

where B_i is the identical bias value for BSs in the ith tier and P_i is the transmit power of BSs in the ith tier. A larger value of B_i implies pushing more users to associate with BSs in the ith tier. Under the above BS association model, the association probability A_k, which is defined as the probability that a user associates with the kth tier, is given by [41]:

$$A_k = \frac{\lambda_{b,k}\left(B_k P_k\right)^\delta}{\sum_i \lambda_{b,i}\left(B_i P_i\right)^\delta}. \tag{5.41}$$

Furthermore, the pdf of the distance D_k conditioned by the user association with the kth tier is given by [41]:

$$f_{D_k}\left(d\right) = \frac{2\pi\lambda_{b,k}}{A_k} d \exp\left(-\pi\frac{\lambda_{b,k}}{A_k}d^2\right). \tag{5.42}$$

Here, we consider two spectrum reuse modes: spectrum-sharing mode and spectrum-partition mode. In the following, we set the time slot length in the ith tier, τ_i, to be inversely proportional to the allocated bandwidth. In spectrum-sharing mode, both tiers reuse the whole bandwidth, W, and the time slot length is set as $\tau_i = Z / W$ with Z being a constant. In spectrum-partition mode, the whole bandwidth is orthogonally partitioned and allocated to the ith tier with a bandwidth $\eta_i W$, satisfying $\sum_i \eta_i = 1$. The time slot length is then set as $\tau_i = Z / (\eta_i W)$ to ensure that the amount of transmitted bits in one time slot is the same for BSs in different tiers, that is, $b = Z \log_2\left(1+\theta\right)$. Under this model, the mean packet delay in radio access links is stated in the following proposition.

Proposition 10: *In the spectrum-sharing mode, the mean packet delay in radio access links of the kth tier is given by:*

$$\overline{T_{ran,k}} = \tau_k \frac{A_k \lambda_u}{\lambda_{b,k}}\left(1 + \frac{A_k}{\lambda_{b,k}}\sum_i \lambda_{b,i} \frac{\delta\theta}{1-\delta}\left(\frac{B_i P_i}{B_k P_k}\right)^\delta \frac{B_k}{B_i} {}_2F_1\left(1,1-\delta;2-\delta;-\theta\frac{B_k}{B_i}\right)\right) \tag{5.43}$$

where ${}_2F_1(\cdot)$ is the Gaussian hypergeometric function.

In the spectrum-partition mode, the mean packet delay in radio access links of the kth tier is given by:

$$\overline{T_{ran,k}} = \tau_k \frac{A_k \lambda_u}{\lambda_{b,k}}(1 + A_k \rho(\alpha,\theta)). \tag{5.44}$$

Proof: The mean packet delay in radio access links is the mean time to successfully transmit a packet from a BS to a user. In the kth tier, it is the product of time slot length τ_k with the mean number of users associating with a BS $\frac{A_k \lambda_u}{\lambda_{b,k}}$ and the mean number of transmissions to successfully transmit a packet $\frac{1}{P_{ss,k}}$, that is:

$$\overline{T_{ran,k}} = \tau_k \frac{A_k \lambda_u}{\lambda_{b,k}} \frac{1}{P_{ss,k}} \tag{5.45}$$

where $p_{ss,k}$ is the success probability in each transmission in the kth tier.

Spectrum-sharing mode: Conditional on the user associating with the kth tier with distance d, the success probability in each transmission is given by [41]:

$$
\begin{aligned}
P_{ss,k}(d) &= \exp\left(-\pi \sum_i \lambda_{b,i} \delta\theta^\delta \left(\frac{P_i}{P_k}\right)^\delta d^2 \int_{\frac{B_i}{B_k}\theta^{-1}}^{\infty} \frac{u^{\delta-1}}{1-u} du \right) \\
&= \exp\left(-\pi d^2 \sum_i \lambda_{b,i} \frac{\delta\theta}{1-\delta}\left(\frac{B_i P_i}{B_k P_k}\right)^\delta \frac{B_k}{B_i}\, {}_2F_1\left(1,1-\delta;2-\delta;-\theta\frac{B_k}{B_i}\right) \right)
\end{aligned} \tag{5.46}
$$

where the second equality holds because the integration can be expressed in terms of the Gaussian hypergeometric function ${}_2F_1(\cdot)$ [42]. Taking expectation with the random distance d with the pdf given by Equation (5.42), the success probability in the radio access link in the kth tier can be obtained as:

$$
\begin{aligned}
P_{ss,k} &= \int_0^\infty P_{ss,k}(d) f_{D_k}(d) dd \\
&= \frac{\lambda_{b,k}/A_k}{\lambda_{b,k}/A_k + \sum_i \lambda_{b,i} \frac{\delta\theta}{1-\delta}\left(\frac{B_i P_i}{B_k P_k}\right)^\delta \frac{B_k}{B_i}\, {}_2F_1\left(1,1-\delta;2-\delta;-\theta\frac{B_k}{B_i}\right)}.
\end{aligned} \tag{5.47}
$$

Substituting Equation (5.47) into Equation (5.45) results in Equation (5.43).

Spectrum-partition mode: In the spectrum-partition mode, the interference in the kth tier only comes from the BSs in that tier. In the interference-limited case, similar to the sub-6 GHz backhaul, given the user's association with the kth tier and the distance between the user and BS d, the success probability is $\exp(-\pi\lambda_{b,k}\rho(\alpha,\theta)d^2)$. Taking expectation with the random distance d with the pdf given by Equation (5.42), the mean success probability in the kth tier is given by:

$$
\begin{aligned}
p_{ss,k} &= \int_0^\infty \exp(-\pi\lambda_{b,k}\rho(\alpha,\theta)d^2) f_{D_k}(d) dd \\
&= \frac{1}{1+A_k\rho(\alpha,\theta)}.
\end{aligned} \tag{5.48}
$$

Similar to the spectrum-sharing mode, substituting Equation (5.48) into Equation (5.45) results in Equation (5.44).

5.5.1.2   Mean Backhaul Delay

Consider the backhaul network model presented in Section 5.4 where the locations of the BSs connecting to gateways with fibre, mmWave, xDSL and sub-6 GHz backhauls are represented as independent thinning of $\Phi_{b,2}$, with probabilities p_f, p_m, p_x and p_s, respectively. The total mean backhaul packet delay is obtained as:

$$\overline{T}_{bh} = p_f \overline{T}_{bh,fb} + p_x \overline{T}_{bh,xd} + p_s \overline{T}_{bh,s6} + p_m \overline{T}_{bh,mm} \tag{5.49}$$

where $\overline{T}_{bh,fb}$ and $\overline{T}_{bh,xd}$ are the mean backhaul delays of fibre and xDSL, which can be obtained via replacing r in Equation (5.14) with r_{fibre} and r_{xDSL}, respectively. $\overline{T}_{bh,s6}$ and $\overline{T}_{bh,mm}$ are the mean backhaul delays of sub-6 GHz and mmWave, which can be obtained from Equations (5.16) and (5.20), respectively.

5.5.1.3   Mean Network Packet Delay

Combining with the mean packet delay in radio access links given by Proposition 10, the mean network packet delay is obtained as:

$$\overline{T} = A_1 \overline{T}_{ran,1} + A_2 \left(\overline{T}_{ran,2} + \frac{A_2 \lambda_u}{\lambda_{b,2}} \overline{T}_{bh} \right) \tag{5.50}$$

where $\overline{T}_{ran,1}$ and $\overline{T}_{ran,2}$ are the mean radio access delay of macro-cell and small-cell tiers, respectively. Note that the coefficient of the backhaul delay comes from the mean number of scheduled users in the backhaul link.

5.5.2   BS Association Policy

Now, we present the BS association problem, which aims to minimize the mean network packet delay. Note from Equation (5.43) that in the spectrum-sharing mode, the bias factor has an effect on both the load and signal quality in each tier. The problem has no simple expression and is not convex in general. For ease of exposition, we focus on the spectrum-partition mode.

The objective of the BS association problem is to find the optimal bias values, B_k^*, which is a one-to-one mapping of the association probability through Equation (5.41). Therefore, we first find the optimal association probability A_k^*, then translate this into B_k^*. Therefore, the BS association problem can be formulated as:

$$\min_{A_k, \eta_k} \quad \overline{T} \tag{5.51a}$$

$$\text{s.t.} \quad \sum_k A_k = 1, A_k \geq 0 \tag{5.51b}$$

$$\sum_k \eta_k = 1, \eta_k \geq 0. \tag{5.51c}$$

Furthermore, $\overline{T_{bh}}$ does not depend on A_k and η_k, which can be considered a system-dependent constant in the optimization problem. Substituting Equation (5.44) into Equation (5.50), the mean network packet delay in Equation (5.51a) can be rewritten as:

$$\overline{T} = \sum_k \tau_k A_k^2 \frac{\lambda_u}{\lambda_{b,k}} (1 + A_k \rho) + A_2^2 \frac{\lambda_u}{\lambda_{b,2}} \overline{T_{bh}}. \tag{5.52}$$

Applying Equation (5.52) and $\tau_k = Z/(\eta_k W)$, the BS association problem in Equation (5.51) can be rewritten as:

$$\min_{A_k, \eta_k} \sum_k A_k^2 \frac{Z/W}{\eta_k} \frac{\lambda_u}{\lambda_{b,k}} (1 + A_k \rho) + A_2^2 \frac{\lambda_u}{\lambda_{b,2}} \overline{T_{bh}} \tag{5.53a}$$

$$\text{s.t.} \quad \sum_k A_k = 1, A_k \geq 0 \tag{5.53b}$$

$$\sum_k \eta_k = 1, \eta_k \geq 0. \tag{5.53c}$$

First, we consider the intuitive solution to set $\eta_k = A_k$, that is, to allocate a fraction of spectrum to the kth tier equal to the association probability [35]. Then, the optimization problem becomes:

$$\min_{A_k, \eta_k} \sum_k \left(A_k^2 \frac{\lambda_u}{\lambda_{b,k}} \left(\frac{Z}{W} \rho + \overline{T_{bh,k}} \right) + A_k \frac{\lambda_u}{\lambda_{b,k}} \frac{Z}{W} \right) \tag{5.54a}$$

$$\text{s.t.} \quad \sum_k A_k = 1, A_k \geq 0 \tag{5.54b}$$

where $\overline{T_{bh,k}}$ are defined for a unified expression, such that $\overline{T_{bh,1}} = 0$ and $\overline{T_{bh,2}} = \overline{T_{bh}}$ are the mean backhaul delays for macro-cell and small-cell BSs, respectively. As the objective function is convex and the constraint is linear, Equation (5.54) is a standard convex optimization problem. Applying the Karush–Kuhn–Tucker (KKT) conditions, the optimal association probabilities can be obtained as:

$$A_k^* = \left(\frac{u\lambda_{b,k}}{2\lambda_u \left(\frac{Z}{W} \rho + \overline{T_{bh,k}} \right)} - \frac{\frac{Z}{W}}{2 \left(\frac{Z}{W} \rho + \overline{T_{bh,k}} \right)} \right)^+ \tag{5.55}$$

where $(x)^+ = \max\{x, 0\}$ and u is the Lagrange multiplier such that $\sum_k A_k = 1$. Therefore, the optimal bias factors can be calculated as [34]:

$$B_k^* = \frac{P_k^{-1}\left(A_k^*/\lambda_{b,k}\right)^{\frac{\alpha}{2}}}{\sum_i P_i^{-1}\left(A_i^*/\lambda_{b,i}\right)^{\frac{\alpha}{2}}}. \tag{5.56}$$

The optimal association probability of a specific tier may be 0 if the BS density of that tier is too small. In that case, little spectrum is allocated to this tier, which makes the radio access delay relatively large and finally forces all users to associate with the other tier. This motivates us to consider the more aggressive spectrum reuse mode, that is, the spectrum-sharing mode, while the association problem in that case is complicated to obtain a general closed-form solution.

In the feasible region where the association probabilities of both tiers are positive, by solving Equations (5.55) and (5.56), the optimal values are obtained as:

$$A_1^* = \frac{\lambda_{b,1}}{\lambda_{b,1} + \dfrac{1}{1+T_{bh}'}\lambda_{b,2}} + \frac{1}{2\rho}\frac{\lambda_{b,1} - \lambda_{b,2}}{\left(1+T_{bh}'\right)\lambda_{b,1} + \lambda_{b,2}} \tag{5.57a}$$

$$A_2^* = \frac{\lambda_{b,2}}{\left(1+T_{bh}'\right)\lambda_{b,1} + \lambda_{b,2}} + \frac{1}{2\rho}\frac{\lambda_{b,2} - \lambda_{b,1}}{\left(1+T_{bh}'\right)\lambda_{b,1} + \lambda_{b,2}} \tag{5.57b}$$

where $T_{bh}' = \dfrac{\overline{T_{bh}}}{\rho Z/W}$. Note from Equations (5.57a) and (5.57b) that the optimal association probability depends on the relative value of the BS densities other than the absolute values. Furthermore, the optimal association probability of the macro-cell tier increases with the mean backhaul delay of the small-cell tier, which is consistent with our intuition.

5.5.3   Numerical Results

In this section, we evaluate the impact of backhaul on mean network packet delay in small-cell networks and the effectiveness of the proposed BS association strategy via numerical results. The densities of users and macro-cell BSs are set to $2 \times 10^{-4}/m^2$ and $10^{-6}/m^2$, respectively. The bandwidth of the radio access network is $W = 40\,MHz$. The probabilities of backhaul deployments are $p_f = 0.4$, $p_x = 0.05$, $p_m = 0.45$ and $p_s = 0.1$. The transmit power ratio of macro-cell and small-cell BSs is $P_1/P_2 = 20$.

5.5.3.1 Effect of Backhaul on Mean Network Packet Delay in Small-cell Networks

The mean network packet delay in small-cell networks under different small-cell BS and gateway densities is presented in Figure 5.5. As observed in Figure 5.5(a), with the small-cell BS density increasing, the backhaul delay remains almost constant, while the radio access delay decreases quickly. When the small-cell BS density is greater than 10^{-5}, the radio access delay is relatively small and the backhaul delay dominates the mean network packet delay. Similarly, as observed in Figure 5.5(b), increasing the gateway density can effectively decrease the backhaul delay, while the

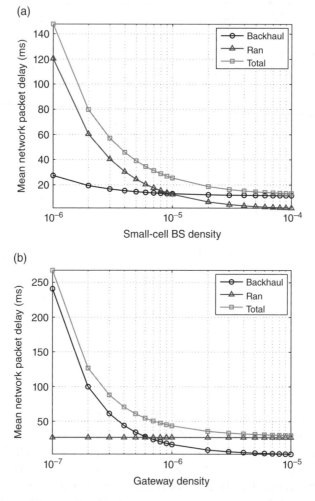

Figure 5.5 Effect of (a) small-cell BS and (b) gateway densities on mean network packet delay. Reproduced from [43] with permission from the IEEE

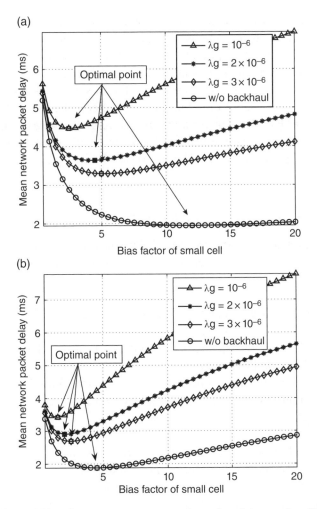

Figure 5.6 Effect of bias factor on mean network packet delay under different spectrum modes: (a) spectrum-partition mode; (b) spectrum-sharing mode. Reproduced from [43] with permission from the IEEE

radio access delay will dominate the mean network packet delay if the gateway density is above a certain threshold (4×10^{-7} in Figure 5.5(b)). This implies that densifying the network via simply deploying a more number of small cell BSs cannot improve system performance indefinitely, and the gateway density in the backhaul network should scale proportionally to the small-cell BS density.

Figure 5.6 shows the effect of biasing on the mean network packet delay in a two-tier cellular network under different backhaul capabilities. The optimal point corresponds to the optimal bias factor, which minimizes the mean network packet delay. First, comparing the two spectrum use modes, the minimum mean network packet delay in spectrum-sharing mode is much smaller as compared to the spectrum-partition mode, as the effect of the bias factor in the spectrum-partition mode is more significant. Specifically, the optimal bias factor that minimizes the mean network packet delay in spectrum-sharing mode is much lower than that in spectrum-partition mode, which means that biasing is not preferable in spectrum-sharing mode due to higher interference for users in the biased region. Secondly, the optimal bias factor increases with the gateway density and will converge to the extreme case without backhaul. This validates that the backhaul links in small-cell networks will deteriorate the offloading gains of small-cell networks with biasing.

5.5.3.2   Comparisons of BS Association Policies

In the following, we compare the proposed BS association strategy with some conventional ones to evaluate its effectiveness. The following three conventional strategies are chosen for comparison:

- No bias, where the user associates with the BS from which it receives the largest average power;
- Distance-based, where the user associates with the nearest BS, no matter whether it is a macro-cell or small-cell BS;
- Bias without backhaul-aware, where the user associates with the BS from which it receives the largest biased average power, using a bias value that minimizes the radio access delay.

The spectrum allocation ratios are all set to be the same as the association ratios. The bias factor for macro cells is set to one. Figures 5.7 and 5.8 present the mean network packet delay, the corresponding association probability and the normalized bias factor under different small-cell BS and gateway densities, respectively.

One can observe from Figure 5.7(a) that the mean network packet delay with the proposed association policy is always the lowest, which validates the effectiveness of the proposed backhaul-aware BS association strategy. Compared with the distance-based and bias-without-backhaul-aware strategies, the mean network packet delay of the proposed strategy decreases much faster as the small-cell BS density increases. This is due to the fact that the effect of backhaul delay becomes gradually more significant with the small-cell BS density and dominates the radio access delay, as shown in Figure 5.5(a). The no bias strategy may even outperform the biased ones without taking backhaul into consideration. This again validates the intuitive idea that conventional biased strategies may mistakenly push more users to the small-cell network (see Figure 5.7(b)), thus deteriorating the delay performance. Figures 5.7(c) and 5.7(b) show that the bias factor decreases with increasing small-cell BS density, and the association probability increases at a diminishing rate. Finally, the association probability of the small-cell tier is even smaller than the case without bias, which implies that users need to be pushed to the macro-cell network.

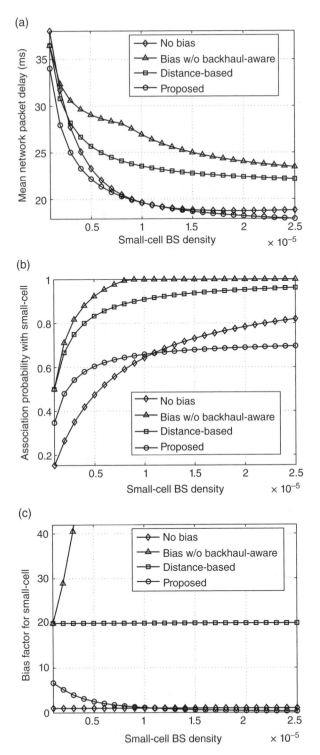

Figure 5.7 Effect of small-cell BS density on BS association. (a) Mean network packet delay; (b) small-cell association probability; (c) normalized small-cell bias factor. Reproduced from [43] with permission from the IEEE

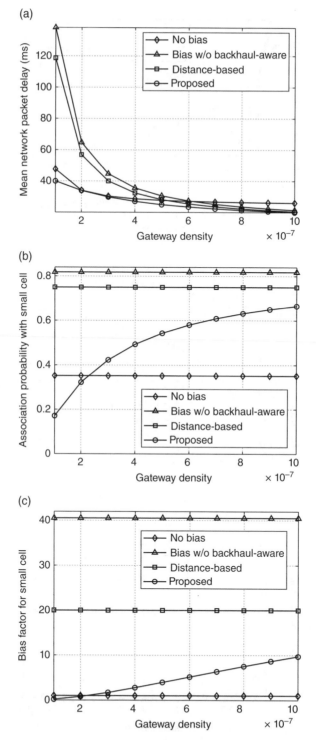

Figure 5.8 Effect of gateway density on BS association. (a) Mean network packet delay; (b) small-cell association probability; (c) normalized small-cell bias factor. Reproduced from [43] with permission from the IEEE

Lastly in Figure 5.8, as the gateway density increases, the mean network packet delay decreases and the performance gap between the proposed strategy and the other two biased strategies becomes smaller. This is because the backhaul delay decreases continuously and the radio access delay dominates, which reduces the decrease in the mean packet delay of biased strategies without taking backhaul into consideration. However, the performance with no bias will become worse with increasing gateway density as compared to the biased strategies, which implies that bias is needed in the case of low backhaul delay. In addition, from Figures 5.8(b) and 5.8(c), since the conventional BS association strategies do not consider backhaul, the gateway density does not influence the bias factor or association probability. In summary, we observe that the bias factor in the association policy should be set carefully by taking backhaul and load into consideration.

5.6 Conclusions

In this chapter, four promising backhaul technologies, namely fibre, xDSL, mmWave and sub-6 GHz, are evaluated in terms of delay performance and cost using a spatial backhaul model. The characteristics of wired and wireless transmissions are then investigated, where the former provides reliable transmission with variable processing delay, whereas the transmissions of the latter are unreliable. Our results show that fibre is the best choice in terms of delay performance. Meanwhile, direct transmission with sub-6 GHz instead of multi-hop transmission with mmWave or xDSL is preferred under a strict transmission deadline, while mmWave is a promising solution for links with short distances. In summary, our proposed model and analysis provide fundamental understanding and guidelines for efficient deployment of backhaul infrastructure in future ultra-dense wireless networks.

Based on our proposed backhaul model, a BS association policy for two-tier cellular networks, which aims to minimize the mean network packet delay taking into account the backhaul, is presented. It is observed that the performance is significantly improved using a backhaul-aware BS association policy compared to using the conventional and backhaul-agnostic ones relying on biasing. Therefore, our proposed analytical framework provides an essential foundation for the joint design of radio access and backhaul links and serves to characterize the ultimate performance limits of network densification.

References

[1] Quek, T. Q. S., de la Roche, G., Guvenc, I. and Kountouris, M. (2013) *Small Cell Networks: Deployment, PHY techniques, and resource allocation*, Cambridge University Press.
[2] Bhushan, N., Li, J., Malladi, D., Gilmore, R., Brenner, D., Damnjanovic, A., Sukhavasi, R., Patel, C. and Geirhofer, S. (2014) Network densification: The dominant theme for wireless evolution into 5G. *IEEE Communications Magazine*, **52**(2), 82–89.

[3] Baldemair, R., Dahlman, E., Fodor, G., Mildh, G., Parkvall, S., Selen, Y., Tullberg, H. and Balachandran, K. (2013) Evolving wireless communications: Addressing the challenges and expectations of the future. *IEEE Vehicular Technology Magazine*, **8**(1), 24–30.

[4] Chia, S., Gasparroni, M. and Brick, P. (2009) The next challenge for cellular networks: Backhaul. *IEEE Microwave*, **10**(5), 54–66.

[5] O3b Networks and Sofrecom (2013) *Why latency matters to mobile backhaul*. White paper.

[6] NGMN Alliance (2007) *Small cell backhaul requirements*. White paper.

[7] Small Cell Forum (2013) *Backhaul technologies for small cells: Use cases, requirements and solutions*. Technical report.

[8] Raza, H. (2013) A brief survey of radio access network backhaul evolution: Part II. *IEEE Communications Magazine*, **51**(5), 170–177.

[9] Tipmongkolsilp, O., Zaghloul, S. and Jukan, A. (2011) The evolution of cellular backhaul technologies: Current issues and future trends. *IEEE Communications Surveys & Tutorials*, **13**(1), 97–113.

[10] Naveh, T. (2009) *Mobile backhaul: Fiber vs. microwave*. White paper.

[11] Bartelt, J., Fettweis, G., Wübben, D., Boldi, M. and Melis, B. (2013) Heterogeneous backhaul for cloud-based mobile networks. *Proceedings of the IEEE Vehicular Technology Conference*, Las Vegas, September.

[12] Singh, S., Mudumbai, R. and Madhow, U. (2011) Interference analysis for highly directional 60-GHz mesh networks: The case for rethinking medium access control. *IEEE/ACM Transactions on Networking*, **19**(5), 1513–1527.

[13] Constantinescu, D. (2005) *Measurements and models of one-way transit time in IP routers*. Blekinge Institute of Technology, Licentiate dissertation.

[14] Holleczek, P., Karch, R., Kleineisel, R., Kraft, S., Reinwand, J. and Venus, V. (2006) Statistical characteristics of active IP one way delay measurements. *Proceedings of IEEE ICNS*, Silicon Valley, July.

[15] Hooghiemstra, G. and van Mieghem, P. (2011) *Delay distribution on fixed Internet paths*. Delft University of Technology, Technical report Ser. 20011020.

[16] Kompella, R., Levchenko, K., Snoeren, A. and Varghese, G. (2012) Router support for fine-grained latency measurements. *IEEE/ACM Transactions on Networking*, **20**(3), 811–824.

[17] Papagiannaki, K., Moon, S., Fraleigh, C., Thiran, P. and Diot, C. (2003) Measurements and analysis of single-hop delay on IP backbone networks. *IEEE Journal on Selected Areas in Communications*, **21**(6), 908–921.

[18] Akdeniz, M., Liu, Y., Samimi, M., Sun, S., Rangan, S., Rappaport, T. and Erkip, E. (2014) Millimeter-wave channel modeling and cellular capacity evaluation. *IEEE Journal on Selected Areas in Communications*, **32**(6), 1164–1179.

[19] Bai, T., Vaze, R. and Heath Jr, R. (2014) Analysis of blockage effects on urban cellular networks. *IEEE Transactions on Wireless Communications*, **13**(9), 5070–5083.

[20] Ghosh, A., Thomas, T., Cudak, M., Ratasuk, R., Moorut, P., Vook, F., Rappaport, T., MacCartney, G., Sun, S. and Nie, S. (2014) Millimeter-wave enhanced local area systems: A high-data-rate approach for future wireless networks. *IEEE Journal on Selected Areas in Communications*, **32**(6), 1152–1163.

[21] Mudumbai, R., Singh, S. and Madhow, U. (2009) Medium access control for 60 GHz outdoor mesh networks with highly directional links. *Proceedings of IEEE INFOCOM*, Rio de Janeiro, April.

[22] Zhang, Q., Yang, C. and Molisch, A. (2013) Downlink base station cooperative transmission under limited-capacity backhaul. *IEEE Transactions on Wireless Communications*, **12**(8), 3746–3759.

[23] Zhou, L. and Yu, W. (2013) Uplink multicell processing with limited backhaul via per-base-station successive interference cancellation. *IEEE Journal on Selected Areas in Communications*, **31**(10), 1981–1993.

[24] Ahmed, A., Markendahl, J. and Cavdar, C. (2014) Interplay between cost, capacity and power consumption in heterogeneous mobile networks. *Proceedings of IEEE ICT*, Lisbon, May.

[25] Ahmed, A., Markendahl, J. and Cavdar, C. (2014) Techno-economics of green mobile networks considering backhauling. *Proceedings of European Wireless*, Barcelona, May.

[26] Paolini, M. (2011) *Crucial economics for mobile data backhaul*. White paper.

[27] Mahloo, M., Monti, P., Chen, J. and Wosinska, L. (2014) Cost modeling of backhaul for mobile networks. *Proceedings of IEEE ICC*, Sydney, Australia, June.

[28] Chen, D., Quek, T. and Kountouris, M. (2015) Backhauling in heterogeneous cellular networks: Modeling and tradeoffs. *IEEE Transactions on Wireless Communications*, **14**(6), 3194–3206.

[29] Suryaprakash, V. and Fettweis, G. (2014) An analysis of backhaul costs of radio access networks using stochastic geometry. *Proceedings of IEEE ICC*, Sydney, Australia, June.

[30] Suryaprakash, V. and Fettweis, G. (2014) Modeling backhaul deployment costs in heterogeneous radio access networks using spatial point processes. *Proceedings of IEEE WiOpt*, Hammamet, Tunisia, May.

[31] Andrews, J., Singh, S., Ye, Q., Lin, X. and Dhillon, H. (2014) An overview of load balancing in HetNets: Old myths and open problems. *IEEE Wireless Communications*, **21**(2), 18–25.

[32] Singh, S. and Andrews, J. (2014) Joint resource partitioning and offloading in heterogeneous cellular networks. *IEEE Transactions on Wireless Communications*, **13**(2), 888–901.

[33] Kim, H., de Veciana, G., Yang, X. and Venkatachalam, M. (2012) Distributed α-optimal user association and cell load balancing in wireless networks. *IEEE/ACM Transactions on Networking*, **20**(1), 177–190.

[34] Bao, W. and Liang, B. (2014) Structured spectrum allocation and user association in heterogeneous cellular networks. *Proceedings of IEEE INFOCOM*, Toronto, May.

[35] Lin, Y. and Yu, W. (2014) Joint spectrum partition and user association in multi-tier heterogeneous networks. *Proceedings of IEEE CISS*, Princeton, NJ, March.

[36] Sadr, S. and Adve, R. (2014) Tier association probability and spectrum partitioning for maximum rate coverage in multi-tier heterogeneous networks. *IEEE Communications Letters*, **18**(10), 1791–1794.

[37] Andrews, J., Baccelli, F. and Ganti, G. (2011) A tractable approach to coverage and rate in cellular networks. *IEEE Transactions on Communications*, **59**(11), 3122–3134.

[38] Xia, P., Jo, H.-S. and Andrews, J. (2012) Fundamentals of inter-cell overhead signaling in heterogeneous cellular networks. *IEEE Journal of Selected Topics in Signal Processing*, **6**(3), 257–269.

[39] Bai, T. and Heath Jr., R. (2015) Coverage and rate analysis for millimeter wave cellular networks. *IEEE Transactions on Wireless Communications*, **14**(2), 1100–1114.

[40] Fiorani, M., Tombaz, S., Monti, P., Casoni, M. and Wosinska, L. (2014) Green backhauling for rural areas. *Proceedings of IFIP ONDM*, Stockholm, May.

[41] Jo, H.-S., Sang, Y. J., Xia, P. and Andrews, J. (2012) Heterogeneous cellular networks with flexible cell association: A comprehensive downlink SINR analysis. *IEEE Transactions on Wireless Communications*, **11**(10), 3484–3495.

[42] Gradshteyn, I. S. and Ryzhik, I. M. (2007) *Table of Integrals, Series, and Products*, 7th edition. Academic Press.

[43] Zhang, G., Quek, T. Q. S., Kountouris, M., Huang, A. and Shan, H. (2016) Fundamentals of heterogeneous backhaul design–analysis and optimization. *IEEE Transactions on Communications*, **64**(2), 876–889.

6

Dynamic Enhanced Inter-cell Interference Coordination Strategy with Quality of Service Guarantees for Heterogeneous Networks

Wei-Sheng Lai,[1] Tsung-Hui Chang,[2] Kuan-Hsuan Yeh[3] and Ta-Sung Lee[1]

[1] *Department of Electrical and Computer Engineering, National Chiao Tung University, Hsinchu, Taiwan*

[2] *School of Science and Engineering, The Chinese University of Hong Kong, Shenzhen, China*

[3] *ASUSTeK Computer Inc., Taipei, Taiwan*

6.1 Introduction

For next generation wireless communication systems, a new set of demands such as high spectral/energy efficiency, enhanced cell average/edge throughput, high peak data rates and low latency are now regarded as the primary system design requirements. The Third Generation Partnership Project (3GPP) has developed several advanced techniques for the Long Term Evolution-Advanced (LTE-A) standard such as coordinated multi-point (CoMP) transmission and reception, enhanced downlink/uplink multiple-input multiple-output (MIMO) technologies, carrier aggregation (CA) and HetNets. With the number of mobile devices increasing in the last few years, exponential data traffic growth is a well-established fact in broadband wireless networks. One of the most promising solutions for solving this growing traffic demand is HetNets [1–4].

Backhauling/Fronthauling for Future Wireless Systems, First Edition.
Edited by Kazi Mohammed Saidul Huq and Jonathan Rodriguez.
© 2017 John Wiley & Sons, Ltd. Published 2017 by John Wiley & Sons, Ltd.

HetNets are deployments of small cells with different coverage radii and macro cells in a single geographic area. The dense deployment of small cells could cause inter-cell interference and reduce the performance gains of HetNets. Various techniques [3–6] have been developed in LTE and LTE-A for tracking inter-cell interference. The inter-cell interference coordination (ICIC) technique can coordinate data transmission and interference of two neighbouring cells in the frequency domain [7, 8]. The eICIC approach is a time domain technique that offloads macro users to small cells [9–11]. The coordination of eICIC muting patterns (referred to as almost blank subframes (ABSs)) is supported via the X2 backhaul interface between different base stations (BSs). In a macro-femto deployment scenario, the power control technique is a simple and effective way to manage the interference without the X2 backhaul interface [4].

In HetNets, the throughput performance of ICIC [12], dynamic selection of users [13] and adaptive bias configuration for eICIC [14] have been evaluated over the last few years. However, the above studies assume conventional or semi-static ABS adaptation in the eICIC mechanism. The static optimization of eICIC parameters cannot respond to dynamic network environments appropriately. Recently, validation of the optimal ABS ratio in different scenarios [15], dynamic ABS ratio optimization [16], the combination of eICIC and self-organizing networks (SONs) [17] and the centralized and distributed solution for fast muting adaptation [18] have been investigated for dynamic eICIC mechanisms. To consider the network load along with user mobility, it is necessary to adopt a dynamic eICIC mechanism in wireless communication systems.

At the same time, in HetNets, mobile users are associated with different serving BSs across different regions over time. Because of the limited spectrum bandwidth, the users need to be served within QoS requirements in the cells. To ensure the users' network performance fulfils the requirements of the International Mobile Telecommunications-Advanced (IMT-A) as defined by the International Telecommunication Union (ITU) [19–21], call admission control (CAC) is an important design factor for the next generation mobile communication networks. To support multimedia services within QoS requirements, many policies have been discussed in previous studies [22, 23], such as complete sharing, complete partition, threshold and Markov decision process (MDP) control. In [23], a control policy based on the MDP was proven to be the optimal choice among the above policies. Several CAC policies with MDP [22] or semi-Markov decision processes (SMDPs) [24–28] have been investigated in wireless cellular networks over the past few years.

In this chapter, we introduce a dynamic eICIC mechanism with QoS requirements. Because of the mobility of wireless subscribers, the load and data traffic are different in each active macro cell and small cell. A conventional static eICIC mechanism cannot ensure that the adaptation of the ABS duty cycle corresponds to dynamic network conditions. Only a dynamic eICIC mechanism is suitable for non-static network traffic. The SMDP model is adopted to support the traffic dynamics in multi-class wireless networks. To determine the optimal CAC policy in small cells, an SMDP-based CAC policy is constructed to satisfy the QoS requirements and efficiently utilize resources in the system. Finally, we evaluate the system performance in different ways by proposing the dynamic interference coordination strategy for eICIC with QoS requirements under different utilities.

The chapter is organized as follows. Section 6.2 focuses on the introduction of the mechanism of eICIC, system network architecture and problem statements. In Section 6.3, we present a joint dynamic eICIC and admission control problem for maximizing sum throughput utility and proportional fairness utility. A proposed strategy based on the modified sum throughput utility and proportional fairness utility is shown in this section. Some numerical results of the proposed algorithm are presented in Section 6.4. Finally, the conclusion of the chapter and several possible future works are given in Section 6.5.

6.2 System Model and Problem Statement

6.2.1 Network Environments

Consider multi-tiered cellular networks consisting of N_k macro-cellular systems and a set of coexisting small-cell systems, as shown in Figure 6.1. Each macro base station (MBS) is separated into three sectors and serves N_l individual sets of user equipment (UEs) that can move independently in each sector. The cellular communication system load varies continuously because of UE mobility and traffic dynamics. To balance the network load from the MBS, N_j small-cell base stations (SBSs) are

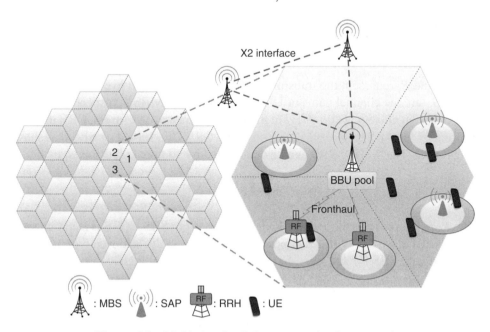

Figure 6.1 Multi-tiered cellular communication network

placed in the macro cell with QoS constraints. Different user motions are modelled by different mobility models that are used to implement the dynamic enhanced inter-cell interference coordination (eICIC) mechanism. In order to ensure a flexible and adaptive system, the channel model and traffic pattern model are also considered, as detailed in Section 6.4 and shown in Table 6.1.

We first consider the downlink transmission of an orthogonal frequency-division multiple access (OFDMA) system. All macro- and small-cell BSs use the same 10 MHz bandwidth at a carrier frequency of 2 GHz. When the time domain eICIC technique is adopted, UEs served by the SBSs can be classified into two groups. UEs within small cell original coverage without bias extension are called 'in-cell' (INC) UEs, while cell-range-extended (CRE) UEs are within the extra cell selection bias (CSB) coverage. Consider INC UE l served by SBS j, as shown in Figure 6.2. To determine the association between the BS and user, each user measures the average reference signal received power (RSRP) for every BS. The serving SBS is determined by choosing the maximum RSRP from SBS j to UE l, given by:

$$\arg\max_{j} RSRP_l^j, \tag{6.1}$$

where $RSRP_l^j$ represents the average RSRP of UE l from SBS j. UEs served by the MBS can also be determined in this way. In the eICIC mechanism, INC users still suffer from interference from the macro cell. The average signal-to-interference-plus-noise ratio (SINR) achieved by INC UE l in the same frequency band can be written as:

$$SINR_{j,l} = \frac{P_j G_{j,l}}{\sum_{j=1, j \neq j'}^{N_j} P_j G_{j,l} + \sum_{k=1}^{N_k} P_k G_{k,l} + z_{j,l}} \tag{6.2}$$

where P_j and P_k represent the transmission power of SBS j and MBS k, respectively, $G_{j,l}$ and $G_{k,l}$ are the channel gains of UE l from SBS j and MBS k, respectively, and $z_{j,l}$ is the Gaussian noise power of UE l in SBS j. In contrast to INC users, CRE users are protected by the ABS from the MBS. Consider a CRE UE i served by SBS j, as shown in Figure 6.2. The serving SBS can be determined by adding a CSB, given by:

$$\arg\max_{j} \left\{ RSRP_i^j + CSB \right\}, \tag{6.3}$$

where $RSRP_i^j$ represents the average RSRP of user i from SBS j. CSB is a non-negative value (in dB) at SBS j. With MBS power muting, the SINR achieved by CRE user i associated with SBS j in the same frequency band can be written as:

$$SINR_{j,i} = \frac{P_i G_{j,i}}{\sum_{j=1, j \neq j'}^{J} P_j G_{j,i} + \sum_{k=2}^{K} P_k G_{k,i} + z_{j,i}}, \tag{6.4}$$

where $G_{j,i}$ and $G_{k,i}$ are the channel gains of UE i from SBS j and MBS k, respectively, and $z_{j,i}$ is the Gaussian noise power of user i in SBS j.

Table 6.1 Parameter settings

Parameter	Value
System	Carrier frequency(f_c): 2 GHz Bandwidth(W_B): 10 MHz
Cell geometry and size	Hexagonal array with ISD, HotSpot: 500 m, sLRB: 1732 m, 3-tier ring, 57 sectors
BS antenna height	MBS: 32 m, UE: 1.5 m
BS antenna horizontal and vertical pattern	MBS: $A_p(\phi, \varphi) = -\min\left\{-\left[A_{pH}(\phi) + A_{pV}(\varphi)\right], A_{att}\right\}$, $A_{pH}(\phi) = -\min\left[12\left(\dfrac{\phi}{\phi_{ab}}\right)^2, A_{att}\right]$ dB, $-180 \le \phi \le 180$, 3-dB beam width of azimuth $\phi_{ab} = 70°$ maximum attenuation $A_{att} = 20$ dB $A_{pV}(\varphi) = -\min\left[12\left(\dfrac{\varphi - \varphi_t}{\varphi_{vb}}\right)^2, A_v\right]$ dB, $\varphi_t = 6°$, 3-dB beam width of vertical $\varphi_{vb} = 10°$, maximum attenuation $A_v = 20$ dB PBS: omnidirectional
Antenna orientation (vertical and azimuth angle)	(0° 30°) for main lobe
Propagation model	MBS: $128.1 + 37.6\log_{10}(d)$ dB, d in km minimum BS–UE separation: 35 m PBS: $140.7 + 36.7\log_{10}(d)$ dB, d in km, minimum BS–UE separation: 10 m
Lognormal shadowing	Correlation distance of shadowing = 50 m MBS: Standard deviation = 8.0 dB PBS: Standard deviation = 10.0 dB
BS shadowing correlation coefficient	Between BSs = 0.5, between sectors of a cell = 1
Mobile noise figure and thermal noise density	9 dB, −174 dbm/Hz
Antenna gains	MBS: 14 dBi, PBS: 5 dBi, UE: 0 dBi
Penetration losses	20 dB
Fast fading model	Doppler $f_d = 70$ Hz (from TR-25.814)
BS maximum power	MBS: 46 dbm PBS: 30 dbm
Number of RBs	50
Number of users	sLRB: 60 HotSpot: 30
Simulation time (T_{sim})	60 seconds
Cell selection bias (CSB)	{0,3,6,9,12,15} dB

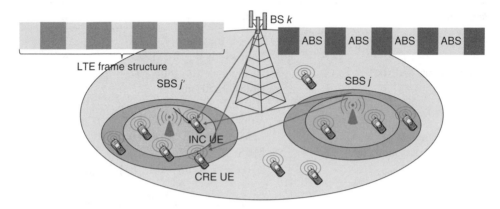

Figure 6.2   INC UEs and CRE UEs in small cells

6.2.2   QoS Constraint

In the QoS condition for HetNets, all additional new users must allow the users in each group to satisfy the QoS requirements or they will be blocked from the group. In order to balance the load from the MBS, we adopt the cell-edge user spectral efficiency and traffic channel link data rates from report ITU-R M.2134 [19] as the minimum required spectral efficiency and resource block (RB) allocation rates, respectively. Assume the users in INC or CRE regions require minimum rates of $R_{\min}^{\mathrm{INC}}$ and $R_{\min}^{\mathrm{CRE}}$, respectively. This gives:

$$R_j^{(s)} = W \cdot s_j^{(s)} \cdot \log_2\left(1 + \mathrm{SINR}_j^{(s)}\right) \geq R_{\min}^{\mathrm{INC}}, \tag{6.5}$$

$$R_j^{(m)} = W \cdot s_j^{(m)} \cdot \log_2\left(1 + \mathrm{SINR}_j^{(m)}\right) \geq R_{\min}^{\mathrm{CRE}}, \tag{6.6}$$

where $\mathrm{SINR}_j^{(s)}$ and $\mathrm{SINR}_j^{(m)}$ are the minimum SINR achieved by the INC and CRE users in small cell j, respectively. The bandwidth of each resource block is W, and the number of RBs for INC users and CRE users is $s_j^{(s)}$ and $s_j^{(m)}$, respectively. In addition to the rate constraints in Equations (6.5) and (6.6), $s_j^{(s)}$ and $s_j^{(m)}$ satisfy:

$$s_j^{(s)} \geq \left\lceil \frac{R_{\mathrm{alloc}}^{\mathrm{INC}}}{W \log_2\left(1 + \mathrm{SINR}_j^{(s)}\right)} \right\rceil, \tag{6.7}$$

$$s_j^{(m)} \geq \left\lceil \frac{R_{\mathrm{alloc}}^{\mathrm{CRE}}}{W \log_2\left(1 + \mathrm{SINR}_j^{(m)}\right)} \right\rceil, \tag{6.8}$$

where $R_{\mathrm{alloc}}^{\mathrm{INC}}$ and $R_{\mathrm{alloc}}^{\mathrm{CRE}}$ are the minimum required RB allocation rates for INC and CRE users, respectively. There are, in total, N RBs in a small cell, and the numbers of allocated RBs for INC and CRE users satisfy:

$$N \geq s_j^{(s)} u_j^{(s)}, \tag{6.9}$$

$$N \geq s_j^{(m)} u_j^{(m)}, \ \forall j = 1, 2, \ldots, N_j, \tag{6.10}$$

where the number of INC and CRE users is $u_j^{(s)}$ and $u_j^{(m)}$, respectively. SBS j chooses the minimum SINR among the INC and CRE users in the cell and calculates the number of allocated resource blocks for each user. SBS j then allocates the same number of resource blocks to other INC and CRE users in the cell, as shown in Figure 6.2. In order to protect the users, the above constraints are used to design the admission control presented in this chapter.

6.2.3   Problem Statements

In this chapter, we consider both the dynamic eICIC and QoS condition for HetNets. In order to evaluate the proposed solution, throughput and fairness are the two metrics used to evaluate the proposed dynamic eICIC mechanism. Throughput and fairness are measured by the sum rate (bps) and the Jain Index in [29], respectively.

We focus on the system utility maximization. A utility function could consist of the system throughput or fairness criteria. In previous work [1], the sum-rate utility was used to measure the sum throughput of the system, so this utility can be modified to evaluate the system performance by adding a QoS constraint. Because of the QoS constraint, the user-blocking probability is imposed on this new utility, called the 'modified sum-rate utility,' which is given as:

$$U_{\Sigma, \mathrm{mod}} = \sum_{\ell=1}^{u^M} (1-\theta) \frac{R_\ell^M}{u^M} + \sum_{j=1}^{N_j} \sum_{i=1}^{u_j^{(m)}} \theta \frac{R_{ji}^{(m)}}{u_j^{(m)}} \left(1 - B_j^{(m)}\right) + \sum_{j=1}^{N_j} \sum_{i=1}^{u_j^{(s)}} (1-\theta) \frac{R_{ji}^{(s)}}{u_j^{(s)}} \left(1 - B_j^{(s)}\right), \tag{6.11}$$

where θ is the duty cycle of the ABS, R_ℓ^M denotes the ith macro user rate in macro m, and $R_{ji}^{(m)}$ and $R_{ji}^{(s)}$ are the rates of the ith CRE user and ith INC user in small cell j, respectively. Further, u^M, $u_j^{(m)}$ and $u_j^{(s)}$ are the numbers of users in macro m, CRE users in small cell j and INC users in small cell j, respectively, and $B_j^{(m)}$ and $B_j^{(p)}$ are the blocking probabilities of CRE and INC users in small cell j, respectively. Moreover, traffic or load condition is decided by the number of present users staying in the small cell. With an approximately equal time over a long period for each user, the proportional fair scheduler creates some fairness for serving throughput to each group of users in the small cell. The modified sum-log utility is given as:

$$U_{\mathrm{log, mod}} = \sum_{\ell=1}^{u^M} \log\left((1-\theta)\frac{R_\ell^M}{u^M}\right) + \sum_{j=1}^{N_j} \sum_{i=1}^{u_j^{(m)}} \log\left(\theta \frac{R_{ji}^{(m)}}{u_j^{(m)}} \left(1 - B_j^{(m)}\right)\right)$$
$$+ \sum_{j=1}^{N_j} \sum_{i=1}^{u_j^{(s)}} \log\left((1-\theta)\frac{R_{ji}^{(p)}}{u_j^{(s)}} \left(1 - B_j^{(s)}\right)\right), \tag{6.12}$$

In addition to the blocking probability of each group of users, the number of RBs is also an important constraint on the SBS. Let the modified utility function be an objective function, and take the ABS duty cycle and the number of RBs as constraints. An optimization problem can then be formulated as:

$$\max_{\theta} U_{\text{mod}}\left(\theta, B_j^{(m)}, B_j^{(s)}\right)$$

$$\text{s.t.} \quad \theta_{\min} \leq \theta \leq \theta_{\max},$$

$$N \geq s_j^{(s)} u_j^{(s)}, \forall j,$$

$$N \geq s_j^{(m)} u_j^{(m)}, \forall j,$$

(6.13)

where the constraint of ABS duty cycle θ can be varied from the minimum value $\theta_{\min}$ to the maximum value $\theta_{\max}$ under the LTE standard. Here, N is the total number of RBs in the small cell. The number of RBs for INC and CRE users in small cell j is $s_j^{(s)}$ and $s_j^{(m)}$, respectively.

6.3 Dynamic Interference Coordination Strategy

6.3.1 SMDP Analysis

For a mobile user, a non-interrupted connection with high quality for audio/video transmission is what he/she cares about. An SMDP model allows multiple class calls that satisfy their optimal QoS traffic parameters in a multi-class environment. Using CAC policies, we can formulate the problem that characterizes the traffic in the environment as an SMDP-based CAC (SMD-CAC) problem [27]. In the SMDP model, the previous state has no relationship with the current system state [30, 31]. Whenever a new call arrives in the small cell, CAC must make decisions based only on the current state. Consider a particular small cell j in the following without loss of generality. A general state in small cell j at epoch t is as follows:

$$\mathbf{x}_j(t) \triangleq \left[u_j^{(s)}(t) \; u_j^{(m)}(t)\right], \forall j,$$

(6.14)

where $u_j^{(s)}(t)$ and $u_j^{(m)}(t)$ denote the numbers of INC and CRE users, respectively, in small cell j at epoch t. We assume one new arrival/departure process in each group at each decision epoch t as an additional process event. The new arrival process is modelled by a Poisson distribution with rate λ, and the departure process uses another Poisson distribution with rate σ. We can then model the two additional event processes as an event process vector. We define the event process vector as follows:

$$\mathbf{e}_{\text{additional}} = \left[e_{\text{INC}}, \; e_{\text{CRE}}\right], e \in \{1, \; 0, \; -1\},$$

(6.15)

where e_{INC} and e_{CRE} denote the event processes for INC and CRE users, respectively. When an event process of the group equals 1 or -1, it respectively represents a connection coming into or departing from that group. Modelling the current state $x_j(t)$ as an original state at epoch t, we can obtain another state by adding the event process vector to the current state. The relationship between the current and next states can be written as:

$$\mathbf{x}'_j(t) = \mathbf{x}_j(t) + \mathbf{e}_{\text{additional}}, \qquad (6.16)$$

where $\mathbf{x}'_j(t)$ denotes the next possible state of the groups of users in small cell j at epoch t. Because the event processes are finite, we must have finite selections on the next possible state. Setting this finite next possible state space as $\mathbf{x}(t)$, a set $\mathbf{A}(\mathbf{x})$ of admissible actions is available for each state in $\mathbf{x}(t)$.

New arrival and departure instances of either INC users or CRE users form the decision epochs of the SMDP. In fact, a CAC needs to be performed only for a new arrival process in either group. We define action $\mathbf{a}$ in small cell j at t as:

$$\mathbf{a}_j(t) = \left[a_j^{(s)}(t) \; a_j^{(m)}(t) \right], \; \forall j, \qquad (6.17)$$

where $a_j^{(s)}(t)$ and $a_j^{(m)}(t)$ denote the actions of INC and CRE users in small cell j at epoch t, respectively. Action $\mathbf{a}$ in small cell j must satisfy the RB constraints in Equations (6.7) and (6.8). When an action of the group equals 1 or 0, an additional new connection request is admitted to or blocked from that group.

If small cell j is in the original state, after action $\mathbf{a}$ is taken but before the next possible state is entered by the system, the expected time $\tau_j(\mathbf{x}, \mathbf{a})$ is given by:

$$\tau_j(\mathbf{x}, \mathbf{a}) = \begin{bmatrix} \lambda_j^{(s)} a_j^{(s)} + \sigma_j^{(s)} u_j^{(s)} + \lambda_j^{(m)} a_j^{(m)} + \sigma_j^{(m)} u_j^{(m)} \\ + \lambda_j^{(s)} a_j^{(s)} \lambda_j^{(m)} a_j^{(m)} + \lambda_j^{(s)} a_j^{(s)} \sigma_j^{(m)} u_j^{(m)} \\ + \sigma_j^{(s)} u_j^{(s)} \lambda_j^{(m)} a_j^{(m)} + \sigma_j^{(s)} u_j^{(s)} \sigma_j^{(m)} u_j^{(m)} \end{bmatrix}^{-1}. \qquad (6.18)$$

The transition probability $p_j(\mathbf{x}, \mathbf{x}', \mathbf{a})$ from the current state $\mathbf{x}_j(t)$ to the next state $\mathbf{x}'_j(t)$ in $\mathbf{X}(t)$ when action $\mathbf{a}$ is taken in small cell j can be written as:

$$p_j(\mathbf{x}, \mathbf{x}', \mathbf{a}) = \begin{cases} \lambda_j^{(m)} a_j^{(m)} \tau_j(\mathbf{x}, \mathbf{a}), & \text{if } \mathbf{x}'_j(t) = \mathbf{x}_j(t) + [0 \; 1] \\ \sigma_j^{(m)} u_j^{(m)} \tau_j(\mathbf{x}, \mathbf{a}), & \text{if } \mathbf{x}'_j(t) = \mathbf{x}_j(t) + [0 \; -1] \\ \lambda_j^{(s)} a_j^{(s)} \tau_j(\mathbf{x}, \mathbf{a}), & \text{if } \mathbf{x}'_j(t) = \mathbf{x}_j(t) + [1 \; 0] \\ \sigma_j^{(s)} u_j^{(s)} \tau_j(\mathbf{x}, \mathbf{a}), & \text{if } \mathbf{x}'_j(t) = \mathbf{x}_j(t) + [-1 \; 0] \\ \lambda_j^{(s)} a_j^{(s)} \lambda_j^{(m)} a_j^{(m)} \tau_j(\mathbf{x}, \mathbf{a}), & \text{if } \mathbf{x}'_j(t) = \mathbf{x}_j(t) + [1 \; 1] \\ \lambda_j^{(s)} a_j^{(s)} \sigma_j^{(m)} u_j^{(m)} \tau_j(\mathbf{x}, \mathbf{a}), & \text{if } \mathbf{x}'_j(t) = \mathbf{x}_j(t) + [1 \; -1] \\ \sigma_j^{(s)} u_j^{(s)} \lambda_j^{(m)} a_j^{(m)} \tau_j(\mathbf{x}, \mathbf{a}), & \text{if } \mathbf{x}'_j(t) = \mathbf{x}_j(t) + [-1 \; 1] \\ \sigma_j^{(s)} u_j^{(s)} \sigma_j^{(m)} u_j^{(m)} \tau_j(\mathbf{x}, \mathbf{a}), & \text{if } \mathbf{x}'_j(t) = \mathbf{x}_j(t) + [-1 \; -1] \\ 0, & \text{otherwise.} \end{cases} \qquad (6.19)$$

6.3.2 Admission Control with a QoS Constraint

In order to provide a better service to small cell users, a QoS condition is imposed in our system. Users in the small cell require different numbers of RBs according to their SINRs. Adding QoS requirements in the small cell means that the user-blocking probability must be considered for the utility function. In the SMDP model, we first minimize the blocking probability in small cell j that can be formulated as an SMD-CAC problem. The SMD-CAC problem is a convex optimization problem which can be solved by a linear programming (LP) approach. The problem is defined as follows:

$$\min_{z_{x,a} \geq 0} \left(\begin{array}{l} \displaystyle\sum_{x \in X} \sum_{a \in A(x)} \left(1 - a_j^{(s)}\right) z_{j,x,a} \tau_j\left(\mathbf{x}, \mathbf{a}\right) \\ \displaystyle + \sum_{x \in X} \sum_{a \in A(x)} \left(1 - a_j^{(m)}\right) z_{j,x,a} \tau_j\left(\mathbf{x}, \mathbf{a}\right) \end{array} \right)$$

$$\text{s.t.} \quad \sum_{a \in A_y} z_{j,y,a} - \sum_{x \in X} \sum_{a \in A(x)} P_{j,xy}\left(\mathbf{a}\right) z_{j,x,a} = 0, \forall \mathbf{y} \in \mathbf{X}(t), j$$

$$\sum_{x \in X} \sum_{a \in A(x)} z_{j,x,a} \tau_j\left(\mathbf{x}, \mathbf{a}\right) = 1, \forall j,$$

$$B_j^{(m)} = \sum_{x \in X} \sum_{a \in A(x)} \left(1 - a_i^{(m)}\right) z_{i,x,a} \tau_i\left(\mathbf{x}, \mathbf{a}\right) \leq 1, \forall j,$$

$$B_j^{(s)} = \sum_{x \in X} \sum_{a \in A(x)} \left(1 - a_j^{(s)}\right) z_{j,x,a} \tau_j\left(\mathbf{x}, \mathbf{a}\right) \leq 1, \forall j,$$

(6.20)

where the first two constraints in Equation (6.20) represent the standard MDP constraints. One is the balance equation and the other describes the fact that the steady-state probabilities must sum to one. In addition, $z_{i,x,a}$ denotes the long-term fraction of choosing action $\mathbf{a}$ when small cell j is in state $\mathbf{x}$, and state $\mathbf{y}$ is a next possible state for small cell j in state $\mathbf{x}$.

Based on the SMD-CAC optimization problem, we propose a dynamic admission strategy to decide how many UEs the small cell should serve, as shown in Algorithm 1. First, all UEs measure their RSRPs from the present active BSs. According to 3GPP Technical Report-36.331 (TR-36.331), the UE measures different averaged RSRPs from different BSs over a fixed period, uses them to determine the serving BS and triggers the measure reports (MRs) to it.

Algorithm 1:   Distributed admission control

1. **Start**
2. Each UE measures its own RSRP using Equation **(6.28)**
3. Identify user cell association using Equation **(6.3)**
4. SBS calculates the number of users and determines how many RBs should be assigned for each using Equations **(6.5)** to **(6.8)**
5. **Solve** the SMD-CAC problem in Equation **(6.20)** and calculate the blocking probability

6. **Calculate** rates $\{R_j\}$ using Equations **(6.5)** and **(6.6)**
7. According to the blocking probability, drop the UE $\hat{j}$ that has the smallest rate in each small cell
8. If the UE $\hat{j}$ is in the INC region, then $U_j^{(s)} \leftarrow U_j^{(s)} \backslash \{\hat{j}\}$, where $U_j^{(s)}$ is the set of UEs in the INC region If the UE $\hat{j}$ is in the CRE region, then $U_j^{(m)} \leftarrow U_j^{(m)} \backslash \{\hat{j}\}$, where $U_j^{(m)}$ is the set of UEs in the CRE region
9. **If** $R_j < R_{\min}, \forall j \in U_j^{(s)} \cap B_j^{(s)} \neq B_j^{(s)*}$ and $R_j < R_{\min}, \forall j \in U_j^{(m)} \cap B_j^{(m)} \neq B_j^{(m)*}$, go to Step 6. Do until $R_j \geq R_{\min}, \forall j \in U_j^{(s)} \cap B_j^{(s)} = B_j^{(s)*}$ and $R_j \geq R_{\min}, \forall j \in U_j^{(m)} \cap B_j^{(m)} = B_j^{(m)*}$
10. **Done**

6.3.3 Joint Dynamic eICIC and Admission Control for Sum Rate Maximization

Mobile users are associated with different BSs across different cell coverages and regions over time. The serving SBS receives MRs from different UEs at the same time. According to the MRs and the number of available RBs, the serving SBS will accept or block newly arrived users from other BSs. To the MBS, traditional interference mitigation appears static in ABS. In other words, the ratio of ABS in all transmission frames is unchanged. The static eICIC technique does not consider dynamically changing network traffic. Previous studies [16, 17] evaluated the performance of dynamic eICIC, but did not consider user rates with QoS conditions in small cells. By introducing the blocking probability of each group of users, the modified sum-rate utility can become an objective function that takes the ABS duty cycle and number of RBs as constraints. The optimization problem can be formulated as:

$$\max_{\theta} U_{\Sigma,\mathrm{mod}} = \sum_{\ell=1}^{u^M}(1-\theta)\frac{R_\ell^M}{u^M} + \sum_{i=1}^{J}\sum_{k=1}^{u_i^{(m)}}\theta\frac{R_{ik}^{(m)}}{u_i^{(m)}}(1-B_i^{(m)}) + \sum_{i=1}^{J}\sum_{k=1}^{u_i^{(p)}}(1-\theta)\frac{R_{ik}^{(p)}}{u_i^{(p)}}(1-B_i^{(p)})$$

s.t. $0 \leq \theta \leq 0.6$ $\qquad\qquad\qquad\qquad\qquad\qquad\qquad$ (6.21)

$\qquad N \geq s_i^{(p)}u_i^{(p)},$

$\qquad N \geq s_i^{(m)}u_i^{(m)}, \forall i = 1,2,\ldots J,$

where the constraint of the ABS duty cycle θ can be varied from the minimum (0) to the maximum (0.6) under the LTE standard. As in Equation (6.13), N is the total number of RBs in the small cell, and the number of RBs for INC and CRE users in small cell j is $s_i^{(s)}$ and $s_i^{(m)}$, respectively. The optimal ABS duty cycle can be derived from convex optimization. Because the objective function of the problem is only a first-order equation of the variable (ABS duty cycle), we can analyse the objective function and predict the system performance by deriving the optimal ABS ratio. From Equation (6.21), it is clear that this modified sum-rate utility function is convex in θ. Because the value of θ can be varied from 0 to 0.6 in the constraint, the optimum of ABS duty cycle θ can be solved by the Karush–Kuhn–Tucker condition, which is given as:

$$\theta^{opt}_{\Sigma,mod} = \begin{cases} 0.6 & \text{if } \sum_{i=1}^{J}\sum_{k=1}^{u_i^{(m)}}\frac{R_{ik}^{(m)}}{u_i^{(m)}}\left(1-B_i^{(m)}\right) > \left(\sum_{\ell=1}^{u^M}\frac{R_\ell^M}{u^M} + \sum_{i=1}^{J}\sum_{k=1}^{u_i^{(p)}}\frac{R_{ik}^{(p)}}{u_i^{(p)}}\left(1-B_i^{(p)}\right)\right), \\ 0 & \text{otherwise.} \end{cases} \quad (6.22)$$

Based on the optimization problem in Equation (6.21), we propose a joint dynamic interference strategy and admission control algorithm to solve this problem. First, the SBS determines the optimal user CAC policy in Algorithm 1, then solves the optimization problem in Equation (6.21) to coordinate the interference between the macro cell and the small cell. This strategy is shown in Algorithm 2.

Algorithm 2: Joint dynamic interference strategy and admission control

```
1. Start (t = 0)
2. Each SBS executes Algorithm 1
3. MBS receives all information from the SBSs by the X2
   backhaul interface
4. Solve the optimization problem in Equation (6.21) such that
   the MBS obtains the optimal ABS duty cycle θ in the
   transmission frames
5. If t < T_sim, go to Step 2. Do until t = T_sim
6. Done
```

6.3.4 Joint Dynamic eICIC and Admission Control for Proportional Fairness Maximization

The modified sum-log utility is a sum of logarithms utility that is clearly a convex function. Again, the number of RBs can be used as a constraint to formulate an optimization problem. Selecting the modified sum-log utility as an objective function and taking the range of ABS duty cycle and number of RBs as constraints, an optimization problem corresponding to Equation (6.13) can be written as:

$$\max_\theta U_{log,mod} = \sum_{\ell=1}^{u^M}\log\left((1-\theta)\frac{R_\ell^M}{u^M}\right) + \sum_{j=1}^{N_j}\sum_{i=1}^{u_j^{(m)}}\log\left(\theta\frac{R_{ji}^{(m)}}{u_j^{(m)}}\left(1-B_j^{(m)}\right)\right)$$
$$+ \sum_{j=1}^{N_j}\sum_{i=1}^{u_j^{(s)}}\log\left((1-\theta)\frac{R_{ji}^{(s)}}{u_j^{(s)}}\left(1-B_j^{(s)}\right)\right) \quad (6.23)$$

$$\text{s.t. } 0 \le \theta \le 0.6,$$
$$N \ge s_j^{(s)}u_j^{(s)}, \forall j,$$
$$N \ge s_j^{(m)}u_j^{(m)}, \forall j,$$

We propose a dynamic interference coordination strategy and admission control algorithm based on the optimization problem in Equation (6.23). First, the UEs measure their RSRPs from all active SBSs and MBSs. According to the 3GPP TR-36.331, a UE measures all different averaged RSRPs from all different BSs. The UE then determines the serving SBS or MBS from all active BSs and triggers the MRs based on the recent averaged RSRP. Second, the SBS executes Algorithm 1 to determine how many UEs can be associated with this SBS. The number of users and allocated RBs are calculated at the SBS. The SBS needs to create the optimal CAC policy and obtain the blocking probability for every group of users. Finally, all information is sent to the MBS from the SBSs by the X2 interface. Using the convex optimization technique, the optimization problem in Equation (6.23) can be solved. The MBS obtains the optimal ABS duty cycle by solving Equation (6.23). The proposed algorithm is the same as Algorithm 2.

We can predict the optimal ABS duty cycle ratio because the objective function of the problem in Equation (6.23) is a first-order equation of the ABS duty cycle. The objective function in Equation (6.23) can be written as:

$$
\begin{aligned}
U_{\log,\text{mod}} &= \sum_{\ell=1}^{u^M} \log\left((1-\theta)\frac{R_\ell^M}{u^M} \right) + \sum_{j=1}^{N_j} \sum_{i=1}^{u_j^{(m)}} \log\left(\theta \frac{R_{ji}^{(m)}}{u_j^{(m)}}\left(1-B_j^{(m)}\right) \right) \\
&\quad + \sum_{j=1}^{N_j} \sum_{i=1}^{u_j^{(s)}} \log\left((1-\theta)\frac{R_{ji}^{(s)}}{u_j^{(s)}}\left(1-B_j^{(s)}\right) \right),
\end{aligned}
\tag{6.24}
$$

$$
= \left(u^M + \sum_{j=1}^{N_j} u_j^{(s)} \right)\log(1-\theta) + \left(\sum_{j=1}^{N_j} u_j^{(m)} \right)\log(\theta) + C_R,
$$

where C_R is given by:

$$
C_R = \sum_{\ell=1}^{u^M} \log\left(\frac{R_\ell^M}{u^M} \right) + \sum_{j=1}^{N_j} \sum_{i=1}^{u_j^{(m)}} \log\left(\frac{R_{ji}^{(m)}}{u_j^{(m)}}\left(1-B_j^{(m)}\right) \right) + \sum_{j=1}^{N_j} \sum_{i=1}^{u_j^{(s)}} \log\left(\frac{R_{ji}^{(s)}}{u_j^{(s)}}\left(1-B_j^{(s)}\right) \right).
\tag{6.25}
$$

Hence, the optimal ABS duty cycle of this modified sum-log utility can be solved by finding the partial derivative of the utility:

$$
\frac{\partial U_{\log,\text{mod}}(\theta)}{\partial \theta}\bigg|_{\theta=\theta_{\log,\text{mod}}^{\text{opt}}} = 0.
\tag{6.26}
$$

The optimal ABS duty cycle is given by:

$$
\theta_{\log,\text{mod}}^{\text{opt}} = \frac{\displaystyle\sum_j^{N_j} u_j^{(m)}}{\displaystyle\sum_j^{N_j} u_j^{(m)} + \sum_j^{N_j} u_j^{(s)} + u^M}.
\tag{6.27}
$$

6.4 Numerical Results

There are nineteen cells that make up three tiers in our simulation model. Every site has three sectors, as shown in Figure 6.3. Sector 1 is called the 'simulation sector'. Four pico-cell base stations (PBSs) are uniformly placed in the simulation sector, and these are located at 0.3 inter-site distances (ISD) from the central MBS. In the simulation, the dynamic interference coordination strategy always runs until the end of the simulation time, T_{sim}.

User mobility is an important feature in our simulation model. As in [16] and the 3GPP Technical Report [32, 33], two models, straightline motion with random bouncing (sLRB) and hot spot (HotSpot) are adopted in our simulations. One models unpredictable user motion and the other models predictable user motion. sLRB can model many people wandering about in a city or large plaza. In the sLRB model, users are located uniformly around the simulation sector. Each user moves in a straight line with a random direction. When a user moves to the edge of the sector, he/she bounces back in any direction. Given a speed of 37.8 km/h, users always change their positions within the simulation sector during the simulation period of T_{sim} s.

HotSpot models an event like a baseball game or dinner at a restaurant. In the HotSpot model, two-thirds of the total number of users initially stay in the small cells (with a distance of less than 50 m). These users move in a random direction in a line. Once a user reaches the edge of the small cell, the user bounces back in a random direction. One-third of the users are distributed uniformly in the simulation sector, according to Scenario 4b in [33]. During the simulation period of T_{sim} seconds, this third go to the nearest small cell before an event starts. They stop in the small cell for an interval (which is the duration of the event) then return to their initial locations.

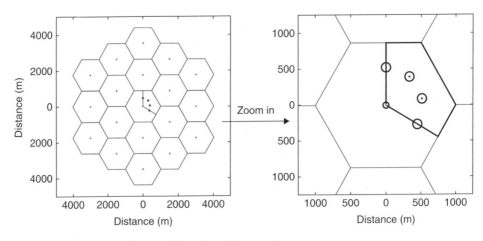

Figure 6.3 Simulation environment set-up

The channel model consists of two parts: the simulation parameter table and radio link monitoring. The radio link monitoring (RLM) procedure is taken from 3GPP TR-36.331. In addition, the simulation parameters are set as in the 3GPP standard [33, 34] and are summarized in Table 6.1. The channel propagation model and antenna pattern are also included in Table 6.1. In the simulation, we assume that the channel estimation, synchronization and instantaneous feedback are perfectly accurate.

A UE has to measure the average RSRP to determine which BS is the serving BS. Because of user motion, the UE re-associates with some different BSs during the simulation. The following is the radio link monitoring process given in [33, 34]:

L1 sampling: sample the 10th RSRP of all SBSs and MBSs every 10 ms.
L3 sampling: take 20 L1 samples from all SBSs and MBSs then compute the average every 200 ms.
L3 filtering: estimate the average RSRP every 200 ms using the following equation:

$$RSRP_{avg}(t) = \frac{3}{4}RSRP_{avg}(t-1) + \frac{1}{4}L3_{avg}. \tag{6.28}$$

We define an A3 event as follows: a UE sends the MRs to the BS that serves it [32]. According to the average RSRPs, a serving BS re-associates its users every 200 ms. First, we know the dynamic eICIC mechanism chooses the extremities of θ automatically from the optimal ABS time evolution pattern. We then compare the static and dynamic eICIC mechanisms. The detailed settings are as follows. The interval, which ranges from 0 to 0.6, is divided into 60 equal points. Each value of the point represents a static eICIC ABS duty cycle. A UE triggers the MRs (A3 event) every 200 ms, and there are 300 sample points of the A3 event during T_{sim} s.

We consider two scenarios and set the CSB to 3 dB for the experiment. In Scenario 1, only a few (e.g., 2–3) UEs are served by the CRE regions of the small cells; in Scenario 2, more (e.g., 7–8) UEs are served by the CRE regions of the small cells. Regardless of whether the dynamic eICIC mechanism selects the minimum or maximum value of θ, the dynamic eICIC sum rate is always better than all instances of the static eICIC sum rates, as shown in Figures 6.4 and 6.5. The advantage of the dynamic eICIC mechanism is clear in these figures.

Finally, we know that the network load allows the ABS ratio to change dynamically. If the small-cell serving coverage is extended, more users might be served by the small cell. When the CSB increases, the maximum value of the ABS duty cycle is chosen more often in the simulation. As the bias increases, more CRE users come into the small cell and are allocated RBs from the SBS. Because the number of RBs increases, the average serving rate of CRE users also increases, as shown in Figure 6.6. If the average serving rate of the CRE users is greater than the sum of the other

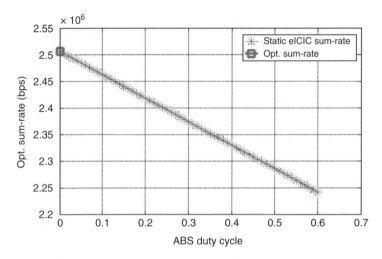

Figure 6.4   Comparison of dynamic and static eICIC sum rates using the sLRB model in Scenario 1

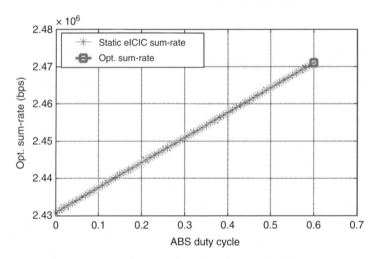

Figure 6.5   Comparison of dynamic and static eICIC sum rates using the sLRB model in Scenario 2

serving users' rates more often, the dynamic eICIC mechanism selects the maximum ABS duty cycle as optimal more often in the simulation.

The system sum rate with QoS requirements is better than the same system without QoS requirements. Because of the rate constraints in Equations (6.7) and (6.8), PBS only serves users satisfying the rate constraints and allocates RBs for them. With QoS requirements and enough RBs, the CSB does not negatively influence the system

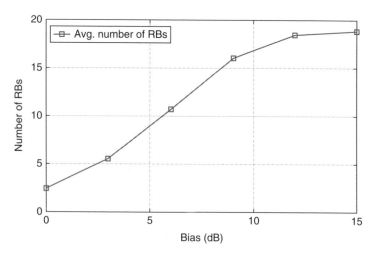

Figure 6.6   Comparison of different biases versus the number of used RBs in the sLRB model

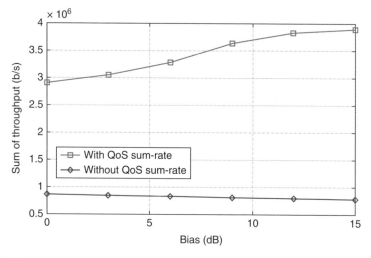

Figure 6.7   Comparison of user sum rates with and without QoS in the sLRB model

performance. Regardless of whether the mobility model in the simulation is sLRB or HotSpot, the performance of both systems has the same trend and low average blocking probability, as shown in Figures 6.7–6.10.

In addition to the comparison of sum rate performances, the user rate average and average small-cell user rate performance are shown by plotting the cumulative distribution function curves in Figures 6.11 and 6.12. With respect to RB assignment, the rate of users with QoS requirements is always better than the rate of users without QoS requirements.

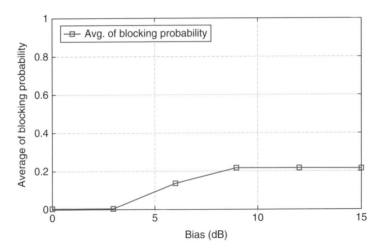

Figure 6.8   User-blocking probability with QoS in the sLRB model

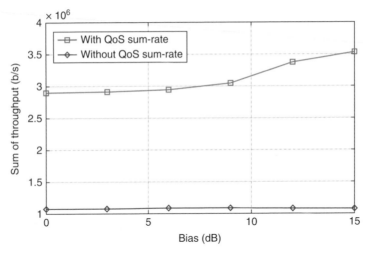

Figure 6.9   Comparison of user sum rates with and without QoS in the HotSpot model

The comparison of the static and dynamic eICIC mechanisms is essential in the simulation. The detailed experiment settings are as follows. The interval, which ranges from 0 to 0.6, is divided into 60 equal points. The same CSB and two scenarios as in Figures 6.4 and 6.5 are adopted. The static eICIC ABS duty cycle value is determined by the point of the interval; there are 60 values of static eICIC sum-log utilities in the experiment. The sum-log utility is a convex optimization and has a maximum value in the interval from 0 to 0.6. Whenever the dynamic eICIC mechanism selects the value for the ABS duty cycle, it always obtains the maximum value of the sum-log utility, as shown in Figures 6.13 and 6.14.

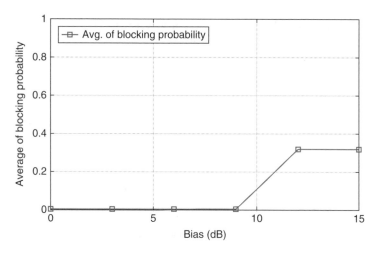

Figure 6.10   User-blocking probability with QoS in the HotSpot model

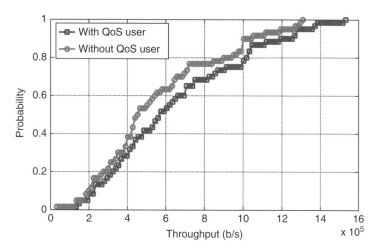

Figure 6.11   Comparison of average user rates with and without QoS in the sLRB model

In order to measure the fairness among users in the system quantitatively, a fairness index must be adopted. The Jain Index is one of the most popular fairness indices used in previous work [29]. It can be written as:

$$\text{Jain Index} = \frac{\left(\sum_{\ell=1}^{L} R_{\ell}\right)^2}{L \cdot \left(\sum_{\ell=1}^{L} R_{\ell}^2\right)}, \tag{6.29}$$

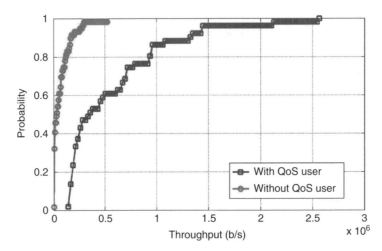

Figure 6.12   Comparison of average small-cell user rates with and without QoS in the sLRB model

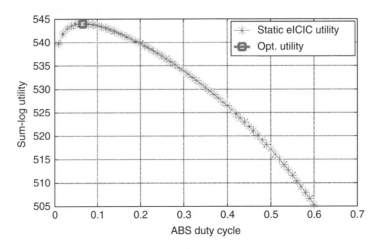

Figure 6.13   Comparison of dynamic and static eICIC sum-log utilities using the sLRB model in Scenario 1

where R_ℓ represents the serving rate of the ℓth user in the system and L is the total number of users in the system. When a single user receives all the resources, the Jain Index is $1/L$. In contrast, the Jain Index is 1 when the same resource allocation is given to each user in the system. Finally, a comparison of the modified sum-rate utility and the modified sum-log utility is essential. In this evaluation, the modified sum-rate utility is used as the throughput metric while the modified sum-log utility is used as the fairness metric. The trade-off between the system sum throughput and fairness is an important factor that needs to be considered. The sum rate performance

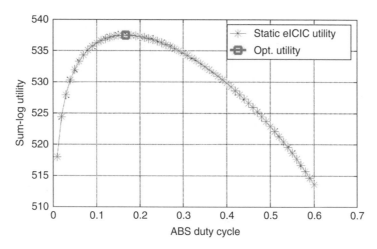

Figure 6.14   Comparison of dynamic and static eICIC sum-log utilities using the sLRB model in Scenario 2

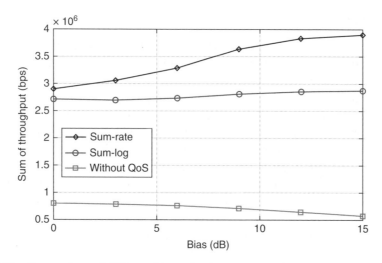

Figure 6.15   Comparison of different biases for the sum rate of the modified sum-rate utility and modified sum-log utility in the sLRB model

of the modified sum-log utility is better in a system with QoS requirements than it is in a system without QoS requirements. However, the Jain Index performance is no longer acceptable because of the minimum requirement of different rates in the system. The sum rate performance of the modified sum-rate utility is better than the modified sum-log utility, but the Jain Index performance of the modified sum-log utility is better than that of the modified sum-rate utility, as shown in Figures 6.15 and 6.16.

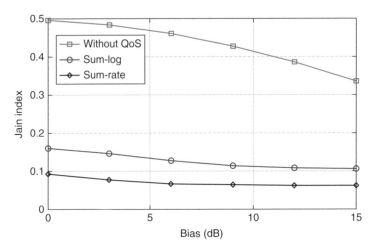

Figure 6.16 Comparison of different biases for the fairness of the modified sum-rate utility and modified sum-log utility in the sLRB model

6.5 Conclusion

For the dynamic interference coordination problem, a dynamic eICIC strategy is a promising solution. This chapter considered QoS requirements such that the small cell must address the cell-association problem to help its users utilize the system resources more efficiently. We thus presented a joint dynamic eICIC mechanism and admission control method to deal with this problem. In contrast to the traditional eICIC mechanism, the proposed method does not add any backhaul requirements because the admission control algorithm handles the admission control in a distributed manner. Depending on the performance evaluation metric, the metric utilities can be categorized into sum-rate utilities and PF utilities. For the modified sum-rate utility, the proposed dynamic interference coordination strategy aims to maximize the system sum throughput under the QoS condition. In the modified PF utility case, the goal of the proposed interference coordination strategy is to maximize the PF utility of user throughput with QoS requirements.

In future work, because further eICIC (FeICIC) [35] will be adopted in 3GPP Release 11 to reduce cell-specific reference signal (CRS) interference in the UE receiver, new techniques for FeICIC are needed to further improve the system performance. Moreover, the coexistence of small cells and WiFi in LTE unlicensed bands (LTE-U) [36] is envisioned to be a future trend for mobile broadband networks.

References

[1] Khandekar, A., Bhushan, N., Tingfang, J. and Vanghi, V. (2010) LTE-Advanced: Heterogeneous networks. In *Wireless Conference (EW)*, pp. 978–982, April.

[2] Damnjanovic, A., Montojo, J., Cho, J., Ji, H., Yang, J. and Zong, P. (2012) UE's role in LTE advanced heterogeneous networks. *IEEE Communications Magazine*, **50**(2), 164–176.

[3] Stanze, O. and Weber, A. (2013) Heterogeneous networks with LTE-Advanced technologies. *Bell Labs Technical Journal*, **18**(1), 41–58.

[4] Acharya, J., Gao, L. and Gaur, S. (2014) *Heterogeneous Networks in LTE-Advanced*, John Wiley & Sons.

[5] Damnjanovic, A., Montojo, J., Wei, Y., Ji, T., Luo, T., Vajapeyam, M., Yoo, T., Song, O. and Malladi, D. (2011) A survey on 3GPP heterogeneous networks. *IEEE Wireless Communications Magazine*, **18**(3), 10–21.

[6] Soret, B., Wang, H., Pedersen, K. I. and Rosa, C. (2013) Multicell cooperation for LTE-advanced heterogeneous networks scenarios. *IEEE Wireless Communications*, **20**(1), 27–34.

[7] Kosta, C., Hunt, B., Quddus, A. and Tafazolli, R. (2013) On interference avoidance through inter-cell interference coordination (ICIC) based on OFDMA mobile systems. *IEEE Communication Surveys Tutorials*, **15**(3), 973–995.

[8] Hamza, A., Khalifa, S., Hamza, H. and Elsayed, K. (2013) A survey on inter-cell interference coordination techniques in OFDMA-based cellular networks. *IEEE Communication Surveys Tutorials*, **15**(4), 1642–1670.

[9] Kamel, M. I. and Elsayed, K. (2012) Performance evaluation of a coordination time-domain eICIC framework based on ABSF in heterogeneous LTE-advanced networks. In *Proceedings of IEEE GLOBECOM*, pp. 5548–5553, December.

[10] López-Pérez, D., Güvenç, I., de la Roche, G., Kountouris, M., Quek, T. Q. and Zhang, J. (2011) Enhanced intercell interference coordination challenges in heterogeneous networks. *IEEE Wireless Communications Magazine*, **18**(3), 22–30.

[11] Nasser, N. and Hassanein, H. (2004) Seamless QoS-aware fair handoff in multimedia wireless networks with optimized revenue. *Proceedings of the Electrical and Computer Engineering Conference*, Canada, pp. 1195–1198, May.

[12] Hong, Y., Lee, N. and Clerckx, B. (2010) System level performance evaluation of inter-cell interference coordination schemes for heterogeneous networks in LTE-A system. In *Proceedings of. IEEE GLOBECOM*, pp. 690–694, December.

[13] Pang, J., Wang, J., Wang, D., Shen, G., Jiang, Q. and Liu, J. (2012) Optimized time-domain resource partitioning for enhanced inter-cell interference coordination in heterogeneous networks. In *Proceedings of IEEE WCNC*, pp. 1613–1617, April.

[14] Tian, P., Tian, H., Zhu, J., Chen, L. and She, X. (2011) An adaptive bias configuration strategy for range extension in LTE-Advanced heterogeneous networks. In *Proceedings of ICCTA*, pp. 336–340, May.

[15] El-Shaer, H. (2012) *Interference management in LTE-Advanced heterogeneous networks using almost blank subframes*. Master's thesis, KTH Vetenskap OCH Konst, Sweden, March 2012.

[16] Vasudevan, S., Pupala, R. and Sivanesan, K. (2013) Dynamic eICIC – A proactive strategy for improving spectral efficiencies of heterogeneous LTE cellular networks by leveraging user mobility and traffic dynamics. *IEEE Transactions on Wireless Communications*, **12**(10), 4956–4969.

[17] Tall, A., Altman, Z. and Altman, E. (2014) Self organizing strategies for enhanced ICIC (eICIC). In *Proceedings of the 12th International Symposium on Modeling and Optimization in Mobile, Ad Hoc, and Wireless Networks, WiOpt*, pp. 318–325, May.

[18] Soret, B. and Pedersen, K. I. (2015) Centralized and Distributed Solutions for Fast Muting Adaptation in LTE-Advanced HetNets. *IEEE Transactions on Vehicular Technology*, **64**(1), 147–158.

[19] Report ITU-R M.2134 (2008) 'Requirements related to technical performance for IMT-Advanced radio interface(s)'.

[20] Parkvall, S., Furuskar, A. and Dahlman, E. (2011) Evolution of LTE toward IMT-Advanced. *IEEE Communications Magazine*, **49**(2), 84–91.

[21] 3GPP (2014) Technical report TR-36.913 V12.0.0, 'Requirements for further advancements for Evolved Universal Terrestrial Radio Access (E-UTRA) (LTE-Advanced),' September.

[22] Ke, K. W., Tsai, C. N., Wu, H. T. and Hsu, C. H. (2008) Adaptive call admission control with dynamic resource reallocation for cell-based multirate wireless systems. In *Proceedings of the IEEE Vehicular Technology Conference*, pp. 2243–2247, May.

[23] Ross, K. W. and Tsang, D. H. K. (1989) Optimal circuit access policies in an ISDN environment: A Markov decision approach. *IEEE Transactions on Communication*, **37**(9), 934–939.

[24] Bartolini, N. and Chlamtac, I. (2002) Call admission control in wireless multimedia networks. In *Proceedings of the IEEE International Symposium on Personal, Indoor and Mobile Radio Communications (PIMRC)*, pp. 285–289, September.

[25] Le, L. B., Hoang, D. T., Niyato, D., Hossain, E. and Kim, D. I. (2012) Joint load balancing and admission control in OFDMA-based femtocell networks. In *Proceedings of the IEEE International Conference on Communications*, pp. 5135–5139, June.

[26] Hong, X., Xiao, Y., Ni, Q. and Li, T. (2006) A connection-level call admission control using genetic algorithm for multi-class multimedia services in wireless networks. *International Journal of Mobile Communications*, **4**(5), 568–580.

[27] Singh, S., Krishnamurthy, V. and Poor, H. (2002) Integrated voice/data call admission control for wireless DS-CDMA systems. *IEEE Transactions on Signal Processing*, **50**(6), 1483–1495.

[28] Choi, J., Kwon, T., Choi, Y. and Naghshineh, M. (2000) Call admission control for multimedia services in mobile cellular networks: A Markov decision approach. In *Proceedings of the 5th IEEE Symposium on Computers and Communications (ISCC)*, pp. 594–599, July.

[29] Jain, R., Chiu, D.-M. and Hawe, W. (1984) 'A quantitative measure of fairness and discrimination for resource allocation in shared computer systems.' DEC Research Report TR-301.

[30] Tijms, H. C. (2003) *A First Course in Stochastic Models*, John Wiley & Sons.

[31] Puterman, M. L. (2005) *Markov Decision Processes: Discrete Stochastic Dynamic Programming*, Wiley-Interscience.

[32] 3GPP (2012) Technical report TR-36.839 V0.7.1, 'Mobility enhancements in heterogeneous networks,' August.

[33] 3GPP (2010) Technical report TR-36.814 V9.0.0, ''Further advancements for E-UTRA physical layer aspects,' March.

[34] 3GPP (2013) Technical report TR-36.133 V11.7.0, ''Requirements for support of radio resource management,' December.

[35] Soret, B., Wang, Y. and Pedersen, K. I. (2012) CRS interference cancellation in heterogeneous networks for LTE-Advanced downlink. In *Proceedings of the IEEE International Conference on Communications (ICC)*, pp. 6797–6801, June.

[36] Abinader Jr, F. M., Almeida, E. P. L., Chaves, F. S., Cavalcante, A. M., Vieira, R. D. *et al.* (2014) Enabling the coexistence of LTE and Wi-Fi in unlicensed bands. *IEEE Communications Magazine*, **22**(11), 54–61.

7

Cell Selection for Joint Optimization of the Radio Access and Backhaul in Heterogeneous Cellular Networks

Antonio De Domenico, Valentin Savin and Dimitri Ktenas
CEA, LETI, MINATEC, Grenoble, France

7.1   Introduction

Dense deployment of small cells is one of the main topics of investigation in 3GPP LTE release 12 [1], which aims to meet the ever-increasing data rate requirements of future generations of wireless communications. In this architecture, classical macro base stations (MeNBs) are complemented with low-power, low-cost nodes to extend the cellular network coverage (both in indoor and outdoor environments) and improve the performance experienced by end users by shortening the distance between mobile terminals and access nodes.

Although heterogeneous networks (HetNets) have captured the attention of the mobile industry since 3GPP LTE release 10, small cells still require further enhancements to enable reliable and energy-efficient operations. In the past, researchers have focused their major efforts on mitigating co-channel interference through enhanced inter-cell interference coordination (eICIC) schemes [2] and coordinated multi-point (CoMP) transmission/reception solutions [3]. Furthermore, adaptive energy-saving mechanisms have been investigated to limit the overall energy consumption in lightly loaded periods by dynamically switching off unnecessary low-power nodes [4].

Backhauling/Fronthauling for Future Wireless Systems, First Edition.
Edited by Kazi Mohammed Saidul Huq and Jonathan Rodriguez.
© 2017 John Wiley & Sons, Ltd. Published 2017 by John Wiley & Sons, Ltd.

Conventionally, the air interface has been considered the most limiting factor in terms of available resources and, therefore, the most important cause of congestion in wireless networks. Such an assumption is correct in traditional macro-cell-based networks, where each cell site has the same backhaul capacity, transmission power and average load. However, in HetNets, this is not true anymore, and the backhaul is one of the main technical challenges for small cells. Small-cell eNBs (SCeNBs) will likely be deployed at about 3–6 m above street level (on street furniture and building facades) to improve the system coverage [5]. However, at these locations, installing fixed broadband access (such as fibre links) for backhaul or line-of-sight (LOS) based microwave links may be too expensive. Hence, in a given area, different small cells will be characterized by heterogeneous backhaul connections with regard to physical design (wired/wireless), capacity, latency and topology. Solutions that jointly optimize the radio access network (RAN) and the backhaul network are required to offer ubiquitous support for high-data-rate wireless services [6].

Here, we investigate backhaul-aware cell-selection mechanisms for HetNets. In the current technology, a piece of user equipment (UE) selects the eNB that is associated with the strongest reference signal received power (RSRP) [7]. Due to the power imbalance between SCeNBs and MeNBs, this solution may prevent UEs from being served by the closest access node. Hence, it leads to limited data rate and increased interference in the uplink, lower battery life at user terminals and limited macro-cell offloading. To deal with these problems, a range expansion technique can be used, where a positive bias is added to the strength of measured signals associated with small cells [8]. This approach implemented jointly with eICIC, which protects range-expanded small-cell UEs, results in improved fairness and network capacity [9, 10]. Nevertheless, some studies have shown that by using large values of range expansion bias, too many UEs may be associated with the same SCeNB, which leads to overload issues [10]. Recently, researchers have investigated joint cell association and resource allocation to achieve fair load balancing in HetNets [11]. To the best of our knowledge, there is a lack of solutions that investigate cell selection in scenarios where the RAN can be constrained by the backhaul network. Ferrus *et al.* have proposed an analytical model to evaluate three cell-selection strategies, which take into account the effect of backhaul [12]. The authors have compared closest cell, radio prioritized and transport network prioritized cell-selection strategies in classical 3G macro-cell networks. However, they do not provide a holistic methodology to optimize the network performance. Furthermore, resource allocation is not considered in their study.

In this chapter we propose load-aware cell-selection mechanisms that jointly take into account the radio access and backhaul characteristics. Our contribution is four-fold. First, in Section 7.2.1, we analyse the relationships amongst the overall network ergodic capacity, the cell load, backhaul constraints and resource-allocation methodologies. We analytically derive the user/cell capacity in the case of round robin, data rate fairness and maximum carrier over interference schemes. Second, in Section 7.2.2, we present the cell-selection problem and we discuss the complexity of obtaining a global optimal solution due its combinatorial nature. Third, in Section 7.3, we present two iterative solutions, named *Evolve* and *Relax*, to solve the cell-association problem with limited complexity. We analytically evaluate their performance and we discuss the implementation costs of the proposed schemes in a 3GPP LTE-Advanced network. Fourth, in Section 7.4, we numerically assess the proposed algorithms by comparing their performance both with the classic scheme based only on the RSRP and with the optimal solution achieved through brute force (BF). Simulation results show that *Evolve* achieves near-optimal performance, leading to very large gains with respect to the other approaches at the cost of limited complexity.

7.2 System Model and Problem Statement

We consider a mobile wireless cellular network in which user terminals and eNBs implement an OFDMA air interface based on 3GPP/LTE downlink (DL) specifications [13]. Coherently with the study on small cell enhancement, which is currently under investigation in 3GPP [14], our research focuses on HetNets where small cells are densely deployed and operate in a dedicated carrier with respect to the macro cell (see Figure 7.1). We also consider the presence of a network controller, which is in charge of the joint optimization of RAN and backhaul functionalities. Finally,

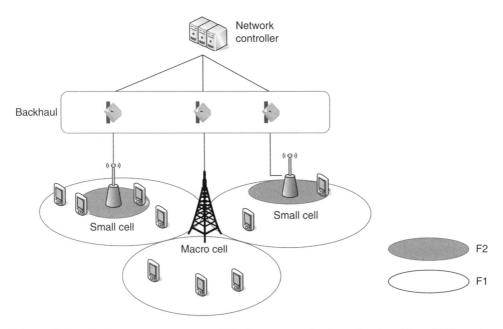

Figure 7.1   The heterogeneous network deployment under investigation. F1 and F2 are the carrier frequencies for the macro layer and the small-cell layer, respectively [14]

we consider non-ideal backhaul that transports information amongst SCeNBs, the controller, MeNBs and the core network.

In the following we denote by U the set of UEs and by S the set of eNBs that offers radio coverage in the HetNet under investigation. UEs are able to be served either by an MeNB operating on F1 or by an SCeNB, which transmits on F2. In conventional networks, UEs keep track of the access nodes whose RSRP is above a given threshold (γ_{th}) and the cell-selection mechanism connects each UE with the eNB associated with the strongest DL signal. According to our model, the link quality between a user u and an eNB s can be modelled through the average SINR as:

$$SINR(u,s) = \frac{P(s) \cdot \Gamma(u,s)}{I(u,s) + \sigma^2} = \frac{P(s) \cdot \Gamma(u,s)}{\sum_{s' \in S \backslash s} P(s') \cdot \mu(s',s) \cdot \Gamma(u,s') + \sigma^2} \tag{7.1}$$

where $P(s)$ is the transmission power at s, $\mu(s',s)$ equals 1 if s and s' operate in the same carrier (otherwise $\mu(s',s)$ equals 0), and $I(u,s)$ is the aggregated interference experienced at u. Moreover, σ^2 is the additive thermal noise power and $\Gamma(u,s)$ is the channel gain which characterizes the radio link, which is defined as:

$$\Gamma(u,s) = \frac{G(u,s)}{L(u,s) \cdot \varsigma(u,s)} \tag{7.2}$$

where $G(u,s)$ is the antenna gain, $L(u,s)$ is the path loss and $\varsigma(u,s)$ is the log-normal shadowing.

Hence, we further denote:

- The bipartite graph, G, with vertices U and S, in which there is an edge between a user u and an eNB s only if u is in the coverage area of s (i.e., $SINR(u, s) \geq \gamma_{th}$);
- $S(u) = \{s \in s \mid (u,s) \in G\}$, the eNBs in the active set of the user u;
- $U(s) = \{u \in U \mid (u,s) \in G\}$, the UEs located in the coverage area of s.

Note that this model is general and may include range expansion analysis by considering different SINR thresholds for MeNBs and SCeNBs.

7.2.1   Joint RAN/BH Capacity

Although a minimum SINR is required at UEs to successfully decode the control channel, the SINR does not provide an intuitive description of the network quality of service (QoS). The data rate experienced at UEs depends on the quality of the link with the associated serving cells; however, each eNB distributes a limited amount of radio resources to multiple users. Hence, user performance is also related to the momentary cell load and the way in which resources are allocated amongst served UEs (the scheduling policy).

Finally, as previously discussed, RAN can be constrained by the backhaul capacity, especially in those cases where the demands of UEs are characterized by very high data rate requirements (like in hotspots). Considering these aspects together in the cell-selection process may be complex, in particular because resource allocation is carried out in a faster time scale than cell association. However, analysing the network ergodic capacity can offer a bound on the overall network performance and provide a key metric to optimize the cell-selection process.

To better understand these relationships, we present in this section an analytical framework that models the achievable rates in HetNets in terms of cell load, backhaul constraints and resource-allocation policy. First, we can define the cell-selection process as follows:

Definition: A cell selection is a mapping $\alpha : U \to S$, such that $\alpha(u) \in S(u), \forall u \in U$. Moreover, $\forall s \in S$, we define $U_\alpha(s) = \{u \in U \mid \alpha(u) = s\} \subset U(s)$ the set of users associated with s. Then, the spectral efficiency (SE) related to the radio link between the user u and its serving eNB $\alpha(u)$ is a logarithmic function of the SINR.

$$\eta(u,\alpha(u)) = \log_2\left(1 + SINR(u,\alpha(u))\right) \tag{7.3}$$

Therefore, the achievable rate at the UE u served by the eNB $\alpha(u)$ can be denoted as:

$$C_\alpha(u) = f_{\alpha(u)} \cdot B_\alpha(u) \cdot \eta(u,\alpha(u)) \tag{7.4}$$

where $B_\alpha(u)$ is the part of the overall bandwidth B allocated to u, which depends on both the radio resource-allocation policy and on the cell load at $\alpha(u)$. Moreover, f_j is a normalization factor, such that:

$$f_{\alpha(u)} = \begin{cases} 1, & \text{if } \displaystyle\sum_{u' \in U_\alpha(\alpha(u))} B_\alpha(u') \cdot \eta(u',\alpha(u)) \leq C^{BH}(\alpha(u)) \\ f_{\alpha(u)} \in (0;1), & \text{otherwise (see below)} \end{cases} \tag{7.5}$$

Equations (7.4) and (7.5) indicate that when the DL transmissions at an eNB $\alpha(u)$ are constrained by the backhaul capacity $C^{BH}(\alpha(u))$, the eNB has to limit the usage of available radio resources, either in terms of allotted bandwidth or SE. In this case, we must have:

$$\sum_{u' \in U_\alpha(\alpha(u))} C_\alpha(u') = C^{BH}(\alpha(u)) \tag{7.6}$$

By plugging Equation (7.4) into Equation (7.6), we can compute $f_{\alpha(u)}$ as:

$$f_{\alpha(u)} = \frac{C^{\text{BH}}(\alpha(u))}{\sum_{u' \in U_\alpha(\alpha(u))} B_\alpha(u') \cdot \eta(u', \alpha(u))} \tag{7.7}$$

Therefore, Equation (7.4) can be rewritten as:

$$C_\alpha(u) = \begin{cases} B_\alpha(u) \cdot \eta(u, \alpha(u)), & \text{if } \sum_{u' \in U_\alpha(\alpha(u))} B_\alpha(u') \cdot \eta(u', \alpha(u)) \leq C^{\text{BH}}(\alpha(u)) \\[2ex] C^{\text{BH}}(\alpha(u)) \cdot \dfrac{B_\alpha(u) \cdot \eta(u, \alpha(u))}{\sum_{u' \in U_\alpha(\alpha(u))} B_\alpha(u') \cdot \eta(u', \alpha(u))}, & \text{otherwise} \end{cases} \tag{7.8}$$

Finally, given the definition of the user achievable data rate, we define the cell capacity as:

$$C_\alpha(s) = \sum_{u \in U_\alpha(s)} C_\alpha(u) \tag{7.9}$$

and the overall network capacity as follows:

$$C(\alpha) = \sum_{s \in S} C_\alpha(s) = \sum_{u \in U} C_\alpha(u) \tag{7.10}$$

To evaluate a bound on the network performance and in line with current 3GPP studies [14], in our analysis we consider full buffer traffic at mobile UEs. However, this model could be extended to other kind of traffic, for instance, by setting the user data rate requirement (given by the application) as a constraint to its achievable rate (see Equations (7.4) and (7.8)).

7.2.1.1 Round Robin (RR)

Let α be a user–cell association. We assume that, by implementing an RR policy, each eNB s, $s \in S$ equally shares the available bandwidth amongst the served users:

$$B_\alpha(u) = \frac{B}{d_\alpha(s)} \quad \forall u \in U_\alpha(s)$$

where $d_\alpha(s) = |U_\alpha(s)|$ is the number of users associated with s. Note that with this approach, UEs characterized by higher SE will experience greater capacity with respect to users with lower SE. Therefore, according to Equation (7.8), for any $u \in U$, we have:

$$C_\alpha(u) = \begin{cases} \dfrac{B}{d_\alpha(s)} \cdot \eta(u, \alpha(u)), & \text{if } \dfrac{B}{d_\alpha(s)} \cdot \displaystyle\sum_{u' \in U_\alpha(\alpha(u))} \eta(u', \alpha(u)) \le C^{\text{BH}}(\alpha(u)) \\ C^{\text{BH}}(\alpha(u)) \cdot \dfrac{\eta(u, \alpha(u))}{\displaystyle\sum_{u' \in U_\alpha(\alpha(u))} \eta(u', \alpha(u))}, & \text{otherwise} \end{cases} \tag{7.11}$$

Moreover, according to Equation (7.9), for any $s \in S$, we have:

$$C_\alpha(s) = \begin{cases} \dfrac{B}{d_\alpha(s)} \cdot \displaystyle\sum_{u \in U_\alpha(s)} \eta(u, s), & \text{if } \dfrac{B}{d_\alpha(s)} \cdot \displaystyle\sum_{u \in U_\alpha(s)} \eta(u, s) \le C^{\text{BH}}(s) \\ C^{\text{BH}}(s), & \text{otherwise} \end{cases} \tag{7.12}$$

7.2.1.2 Data Rate Fairness (DRF)

Here we assume that each eNB equally shares the available capacity amongst the served users. Hence, $\forall (u, u') \in U_\alpha(s)$, we have:

$$B_\alpha(u) \cdot \eta(u, s) = B_\alpha(u') \cdot \eta(u', s)$$

Then, applying

$$B = \sum_{u' \in U_\alpha(s)} B_\alpha(u')$$

we can model the allocated bandwidth per user as follows:

$$B_\alpha(u) = \frac{B}{\eta(u, s) \cdot \displaystyle\sum_{u' \in U_\alpha(s)} \frac{1}{\eta(u', s)}}$$

Hence, according to Equation (7.8), for any $u \in U$, we have:

$$
C_\alpha(u) = \begin{cases}
\dfrac{B}{\sum_{u' \in U_\alpha(\alpha(u))} \dfrac{1}{\eta(u', \alpha(u))}}, & \text{if } \dfrac{B \cdot d_\alpha(s)}{\sum_{u' \in U_\alpha(\alpha(u))} \dfrac{1}{\eta(u', \alpha(u))}} \leq C^{\mathrm{BH}}(\alpha(u)) \\[20pt]
\dfrac{C^{\mathrm{BH}}(\alpha(u))}{d_\alpha(s)}, & \text{otherwise}
\end{cases}
\tag{7.13}
$$

Moreover, according to Equation (7.9), for any $s \in S$, we have:

$$
C_\alpha(s) = \begin{cases}
\dfrac{B \cdot d_\alpha(s)}{\sum_{u' \in U_\alpha(s)} \dfrac{1}{\eta(u, s)}}, & \text{if } \dfrac{B \cdot d_\alpha(s)}{\sum_{u' \in U_\alpha(s)} \dfrac{1}{\eta(u,s)}} \leq C^{\mathrm{BH}}(\alpha(u)) \\[20pt]
C^{\mathrm{BH}}(s), & \text{otherwise}
\end{cases}
\tag{7.14}
$$

7.2.1.3 Max C/I (MCI)

Here we assume that each eNB s allocates more bandwidth to those users that are characterized by greater spectral efficiency. Hence, we have:

$$
B_\alpha(u) = B \cdot \frac{\eta(u, s)}{\sum_{u' \in U_\alpha(s)} \eta(u', s)} \quad \forall \, u \in U_\alpha(s).
$$

According to Equation (7.8), for any $u \in U$, we have:

$$
C_\alpha(u) = \begin{cases}
B \cdot \dfrac{\eta(u, s)^2}{\sum_{u' \in U_\alpha(\alpha(u))} \eta(u', s)}, & \text{if } B \cdot \dfrac{\sum_{u' \in U_\alpha(\alpha(u))} \eta(u', \alpha(u))^2}{\sum_{u' \in U_\alpha(\alpha(u))} \eta(u', \alpha(u))} \leq C^{\mathrm{BH}}(\alpha(u)) \\[20pt]
C^{\mathrm{BH}}(\alpha(u)) \cdot \dfrac{\eta(u, s)^2}{\sum_{u' \in U_\alpha(\alpha(u))} \eta(u', s)^2}, & \text{otherwise}
\end{cases}
\tag{7.15}
$$

Finally, according to Equation (7.9), for any $s \in S$, we have:

$$
C_\alpha(s) = \begin{cases}
B \cdot \dfrac{\sum_{u' \in U_\alpha(s)} \eta(u', s)^2}{\sum_{u' \in U_\alpha(s)} \eta(u', s)}, & \text{if } B \cdot \dfrac{\sum_{u' \in U_\alpha(s)} \eta(u', s)^2}{\sum_{u' \in U_\alpha(s)} \eta(u', s)} \leq C^{\mathrm{BH}}(s) \\[20pt]
C^{\mathrm{BH}}(s), & \text{otherwise}
\end{cases}
\tag{7.16}
$$

7.2.2 Problem Statement

In this chapter, we aim to find the association amongst UEs and eNBs $\alpha*$ that maximizes the overall network capacity. The optimization problem can be simply expressed as:

$$\text{Find } \alpha^* = \text{argmax}_\alpha C(\alpha)$$

At first glance, this combinatorial optimization problem may seem similar to a multiple knapsack problem [15], in which N items (the UEs) have to be associated with M knapsacks (the eNBs), each one of which has a limited weight (the capacity of the corresponding backhaul link), so as to maximize the profit (the overall capacity of the wireless network). In reality, our optimization problem is even more general than the multiple knapsack problem, since the UEs do not have *a priori* weight and profit, but these values are dependent on the association itself and on the resource allocation. Indeed, for each association, α, each user u contributes to the weight of $\alpha(u)$ and to the value of the total profit $C(\alpha)$ according to the quality of the link $(u, \alpha(u))$ and the resource allocation at $\alpha(u)$. Since the knapsack problem is NP-complete, we expect our optimization problem to be so as well, although a formal proof of such a result is beyond the scope of this work. BF algorithms, which evaluate all possible solutions and select the best one might be used to solve simple combinatorial problems. According to our model, the size of the set V, which represents all the feasible solutions, can be computed as:

$$|V| = \prod_{u=1}^{U} |S(u)| \tag{7.17}$$

Therefore, even in moderately dense deployment scenarios, computational/memory costs may prevent us finding an optimal solution by using BF. Henceforth, in the next section, we propose and investigate two iterative algorithms characterized by a limited complexity and designed to improve the overall network capacity by optimizing the cell-selection process.

7.3 Proposed Solutions

In this section, we present two centralized algorithms to manage the cell-selection problem in a near-optimal way. These algorithms require information exchange amongst eNBs and the network controller. In Section 7.3.3 we will discuss the practical implementation of the proposed schemes and in Section 7.4 we will assess their performance through numerical simulations.

7.3.1 Evolve

The first algorithm starts from a given simple solution of the cell-selection problem and evolves towards a more beneficial association. At each iteration, *Evolve* calculates and evaluates each possible change in the current association and then selects the strategy which increases the overall network capacity the most. The algorithm stops after a limited number of iterations, when the achievable gain becomes less than a small, non-negative value ϵ.

(0)   Initialization step

- Let α be the state-of-the-art user assignment that associates to each user u the eNB s maximizing $SINR(u, s)$, that is $\alpha(u) = \text{argmax}_{s \in S} SINR(u, s)$.
- For all $s \in S$, compute $C_\alpha(s)$ according to the used scheduler [cf. Equations (7.12), (7.14) and (7.16)].
- For all $(u, s) \in G$, compute $X_\alpha(u, s)$, which measures the new capacity at the eNB s whether we change the association α by associating (respectively, de-associating) the user u to (respectively, from) s:

$$
X_\alpha(u,s) = \begin{cases}
0, & \text{if } \alpha(u) = s \text{ and } d_\alpha(s) = 1 \\
D_\alpha(s)^{\ominus u}, & \text{if } \alpha(u) = s \text{ and } d_\alpha(s) > 1 \text{ and } D_\alpha(s)^{\ominus u} < C^{BH}(s) \\
D_\alpha(s)^{\oplus u}, & \text{if } \alpha(u) \neq s \text{ and } D_\alpha(s)^{\oplus u} < C^{BH}(s) \\
C^{BH}(s), & \text{otherwise}
\end{cases}
$$

The values of $D_\alpha(s)^{\ominus u}$ and $D_\alpha(s)^{\oplus u}$ with respect to the different resource-allocation policies are shown in Table 7.1.

- For all $(u, s) \in G$, compute the gain $\Delta_\alpha(u, s)$ due to the possible reassignments of the user u from the eNB $\alpha(u)$ to the eNB s:

$$
\Delta_\alpha(u,s) = \begin{cases}
X_\alpha(u,s) + X_\alpha(u, \alpha(u)) - C_\alpha(s) - C_\alpha(\alpha(u)), & \text{if } \alpha(u) \neq s \\
0, & \text{otherwise}
\end{cases}
$$

Table 7.1 $D_\alpha(s)^{\ominus u}$ and $D_\alpha(s)^{\oplus u}$ with respect to different resource-allocation policies

	Round Robin	Data Rate Fairness	Max C/I
$D_\alpha(s)^{\ominus u}$	$\dfrac{B}{d_\alpha(s)-1} \cdot \sum\limits_{u' \in U_\alpha(s)\backslash u} \eta(u',s)$	$\dfrac{B \cdot (d_\alpha(s)-1)}{\sum\limits_{u' \in U_\alpha(s)\backslash u} \frac{1}{\eta(u',s)}}$	$B \cdot \dfrac{\sum\limits_{u' \in U_\alpha(s)\backslash u} \eta(u',s)^2}{\sum\limits_{u' \in U_\alpha(s)\backslash u} \eta(u',s)}$
$D_\alpha(s)^{\oplus u}$	$\dfrac{B}{d_\alpha(s)+1} \cdot \sum\limits_{u' \in U_\alpha(s)\cup u} \eta(u',s)$	$\dfrac{B \cdot (d_\alpha(s)+1)}{\sum\limits_{u' \in U_\alpha(s)\cup u} \frac{1}{\eta(u',s)}}$	$B \cdot \dfrac{\sum\limits_{u' \in U_\alpha(s)\cup u} \eta(u',s)^2}{\sum\limits_{u' \in U_\alpha(s)\cup u} \eta(u',s)}$

(1) One-user reassignment step

- Find $(u_0, s_0) = \text{argmax}_{(u,s) \in G} \Delta_\alpha(u, s)$
  Note that $\Delta_\alpha(u, s) \geq 0$, since $\Delta_\alpha(u, \alpha(u)) = 0$, $\forall u \in U$.
- If $\Delta_\alpha(u_0, s_0) \leq \epsilon$ exit (the algorithm outputs the current user assignment α).
- Define $s_* = \alpha(u_0)$ (hence $s_* \neq s_0$).
- Define a new user assignment α_0 by:

$$\alpha_0(u) = \begin{cases} \alpha(u), & \text{if } u \neq u_0 \\ s_0, & \text{if } u = u_0 \end{cases}$$

(2) Metric-update step

For all $s \in S$,

$$C_{\alpha_0}(s) = \begin{cases} C_\alpha(s), & \text{if } s \neq s_* \text{ and } s \neq s_0 \\ X_\alpha(u, s), & \text{if } s = s_* \text{ or } s \neq s_0 \end{cases}$$

- Set $X_{\alpha_0}(u, s) = X_\alpha(u, s)$, for all $s \in S \backslash \{s_*, s_0\}$ and $u \in U(s)$;
  Compute $X_{\alpha_0}(u, s)$ for $s \in \{s_*, s_0\}$ and $u \in U(s)$.
- Set $\Delta_{\alpha_0}(u, s) = \Delta_\alpha(u, s)$, for all $s \in S \backslash \{s_*, s_0\}$ and $u \in U(s)$;
  Compute $\Delta_{\alpha_0}(u, s)$ for $s \in \{s_*, s_0\}$ and $u \in U(s)$.
- Set $\alpha = \alpha_0$, then go to **Step (1)**.

The proposed framework for backhaul-aware cell selection is illustrated in Figure 7.2.

Proposition: In *Evolve*, the value of $C(\alpha)$ is improved at each new iteration. Hence, the algorithm converges when it is no longer possible to improve the value of $C(\alpha)$ by a new reassignment of one single user.

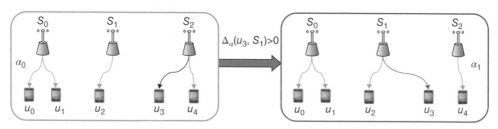

Figure 7.2 *Evolve* paradigm for managing user–eNB association

Proof: Let α be the current user assignment, possibly after some number of iterations. Let α_0 be the new reassignment, computed at Step (1). Then:

$$C(\alpha_0) - C(\alpha) = \sum_{s \in S} C_{\alpha_0}(s) - \sum_{s \in S} C_\alpha(s) = C_{\alpha_0}(s_*) + C_{\alpha_0}(s_0) - C_\alpha(s_*) - C_\alpha(s_0)$$
$$= X_\alpha(u_0, s_*) + X_\alpha(u_0, s_0) - C_\alpha(s_*) - C_\alpha(s_0) = \Delta_\alpha(u_0, s_0) \geq \epsilon$$

In particular, *Evolve* can guarantee at least the same performance as the SINR-based approach.

The proposed algorithm can be further analysed in terms of trellis and the Viterbi algorithm; let's consider the trellis whose state nodes and transitions are defined as follows:

- At time $t = 0$, there is one single state node, corresponding to some user assignment, α. The weight of α is, by definition, $C(\alpha)$.
- State nodes at time $t + 1$ correspond to assignments α'' that differ from some assignment α' at time t in at most one user. A transition from α' to α'' has a weight of $C(\alpha'') - C(\alpha')$.
- We shall further assume that the depth of the trellis (the number of time steps) is equal to the number of users.

Hence, the optimal assignment α^* (maximizing $C(\alpha)$) can be found by determining the maximum-weight path in the trellis, which can be achieved by using the Viterbi algorithm. The proposed algorithm explores a reduced number of paths in the trellis as a trade-off between complexity and optimality of the solution.

7.3.2 Relax

The second proposed algorithm starts by relaxing the constraint that forces each UE to select a single eNB. Let's consider the bipartite graph G defined in Section 7.2. This graph contains all possible solutions of the association problem (see Figure 7.3). *Relax* acts on this graph and at each iteration it first evaluates the value of each edge present in the graph (in the sense of the network capacity); second, it finds and eliminates the edge that corresponds to the less valuable connection; third, *Relax* updates the values on the remaining edges. Finally, the algorithm ends when it reaches an admissible solution, where each UE is associated only with one eNB.

Let U' be the set of UEs that are in the coverage area of at least two eNBs, and let G' be the subgraph of G induced by U'. We denote by $d_G(s)$ the number of edges incident to s. Note that the *Relax* algorithm modifies the graph G by iteratively removing edges until each UE is connected to one single eNB. Each time an edge is removed from G, we also assume that $d_G(s)$, U' and G' are updated accordingly.

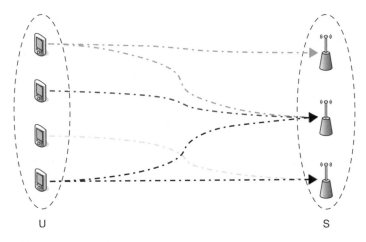

Figure 7.3   The bipartite graph G used to model the cell-association problem

(0) Initialization step
- For all $u' \in U'$, for all $s \in S(u')$,
  - Compute $X_G(u',s)$, which measures the new capacity at the eNB s when we de-associate the user u'.

$$X_G\left(u',s\right)=\begin{cases}0, & \text{if } d_G\left(s\right)=1 \\ D_\alpha\left(s\right)^{\ominus u'}, & \text{if } d_G\left(s\right)>1 \text{ and } D_\alpha\left(s\right)^{\ominus u'}<C^{\text{BH}}\left(s\right) \\ C^{\text{BH}}\left(s\right), \text{otherwise}\end{cases}$$

   where $D_\alpha(s)^{\pounds u'}$ is defined in Table 7.1 with $d_\alpha(s)$ and $U_\alpha(s)$ being replaced by $d_G(s)$ and $U(s)$, respectively.
  - Set $(u',s)=C(s)-X_G(u',s)$, where $C(s)$ is the capacity at the eNB s computed according to the used scheduler [cf. Equations (7.12), (7.14) and (7.16)].

(1) Edge elimination step
- If U' is empty exit.
- Find $(u_0,s_0)=\text{argmin}_{(u',s)\in G'}\, V(u',s)$

(2) Metric-update step
- Remove the edge (u_0,s_0) from G (note that U' and G' are updated accordingly).
- Update $C(s_0)$.
- For any $u'\in U'(s_0)$, update $X_G(u',s_0)$ and $V(u',s_0)$, then go to **Step (1)**.

When *Relax* stops, each user u is associated with a unique eNB, say s_u. We define the user association by $\alpha(u) = s_u$ for all $u \in U$. It is important to underline that this approach cannot guarantee that the overall network capacity is improved at each iteration. Nevertheless, the *Relax* scheme is characterized by a fixed number of iterations, given by:

$$I = \sum_{u \in U}(|S(u)| - 1)$$

which leads to a much faster convergence with respect to the BF algorithm (see Equation (7.17)). For instance, in the example described in Figure 7.3, the proposed algorithm stops after two iterations, since two edges must be removed from the graph.

7.3.3   *Practical Implementation of the Proposed Algorithms*

From the system-level perspective, our algorithms are fully based on interfaces, functions and messages that are already standardized, which results in limited complexity. Both *Evolve* and *Relax* can be implemented through the mobility load balancing (MLB) function, which has been defined in the framework of self-organizing networks (SONs) to improve the LTE performance through coordinated traffic steering [16]. MLB is based on the exchange of information about load level and available capacity amongst neighbouring cells through the X2 interface. Based on these reports, a given algorithm (such as *Evolve*) decides the momentary optimal association amongst UEs and eNBs.

According to the output of the algorithm, cell reselection and handover functions are executed to shift idle and connected UEs to the target eNBs. To reduce complexity and system overhead, the periodicity of reporting can be requested only in the range of 1 to 10s [17]. Hence, fast adaptation (in the scale of milliseconds) to mobility, cell load and channel conditions is not feasible, only average measurements are required, and there are not stringent constraints in terms of latency. Furthermore, in this chapter, we propose to centrally execute the MLB and the associated algorithm to reduce the signalling overhead and permit a more efficient usage of the available resources. Figure 7.4 shows the required message passing by the proposed algorithms. The process is initiated through the MLB trigger that is sent to the network controller by an eNB that is currently overloaded. Then, the controller requests from the overloaded eNB and its neighbouring eNBs measurements on the experienced load, the SINR measured on the radio links, the capacity of the backhaul, etc. By using the received inputs, the proposed algorithm can be implemented and the novel optimal association can be transferred to the set of eNBs to be executed. To enable the successful implementation of the proposed approach, it is also important to limit the additional computational complexity. This cost is related to the number of iterations required by our algorithms to converge. In Sections 7.3.1 and 7.3.2, we have seen that both the two

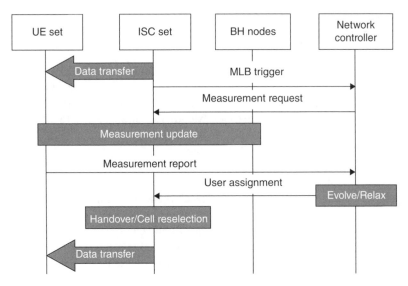

Figure 7.4   The required message passing in the proposed algorithms

proposed algorithms converge in a limited number of iterations. However, we cannot calculate *a priori* the number of iterations needed for *Evolve*. When very low complexity is necessary, the stopping parameter ϵ (see Step 1 in Section 7.3.1) can be adjusted to reduce the complexity of *Evolve* at the cost of lower performance. In the following section, we will show, through numerical results, that our approach needs a limited number of iterations in a realistic scenario.

7.4   Simulation Results

In this section, we assess the effectiveness of the proposed *Evolve* and *Relax* algorithms by comparing their performance with respect to the optimal solution, obtained through the BF algorithm, and the classical approach where each UE selects the eNB associated with the strongest RSRP. Here we assume that SCeNBs form 3×3 grids located inside the macro cell; moreover, two-thirds of the overall UEs are distributed inside the small-cell grids and the remaining UEs are uniformly located in the macro-cell area [14]. Key simulation parameters are detailed in Table 7.2.

The results are averaged over 10^3 independent runs. At the beginning of each run, the clusters of SCeNBs and UEs are randomly deployed in the macro-cell area. In our simulations, UEs include in their active set those eNBs associated with an SINR greater than γ_{th} equal to $-3\,dB$, and the stopping parameter ϵ equals 0. As already mentioned in Section 7.2.1, we consider full buffer traffic at mobile UEs. Finally, the user SE is upper limited to 12 bit/s/Hz (η_{max}) to fairly evaluate the impact of the RAN and backhaul on the overall network performance.

Table 7.2   Main simulation parameters

Parameter	Value	Parameter	Value
Cellular layout	Hexagonal grid	Carrier frequency	2.0 GHz (macro cell) 3.5 GHz (small cell)
Inter-site distance	500 m	Carrier bandwidth	10.0 MHz
Macro sites	19	MeNB Tx power	46 dBm
Macro sector/site	3	MeNB max. antenna gain	13 dBi
SCeNBs/macro sector	9	SCeNB Tx power	30 dBm
UE dropping	2/3 UEs within the clusters	SCeNB antenna gain	5 dBi
Min. distance SC–UE	10 m	MeNB antenna pattern	2D three-sectorized
Min. distance MeNB–UE	35 m	SC antenna pattern	Omnidirectional
Min. distance MeNB–SC	75 m	Shadowing distribution	Log-normal
Distance SC–SC	40 m	Macro/SC LOS probability	See Table A.2.1.1.2-3 [18]
Macro cell path loss	ITU UMa (Table B.1.2.1-1 [18])	Small cell path loss	ITU Umi (Table B.1.2.1-1 [18])
Backhaul capacity for DL data transmission	40/80/120 Mbit/s	Thermal noise density	$N_0 = -174$ dBm/Hz

First, we aim to assess the performance of *Evolve*, *Relax* and the SINR-based algorithms with respect to the optimal solution. Due to the high complexity of the BF algorithm, we consider a light-deployment scenario, composed of an MeNB, a cluster of nine small cells and 20 UEs located in the central macro-cell site. However, eNBs located in the surrounding sites are only used to model inter-cell interference. Figure 7.5 shows the cumulative distribution function (CDF) of the network capacity $C(\alpha)$ achieved with the MCI scheduling policy with respect to different backhaul constraints. Plots using open squares, open circles, filled circles and filled squares, respectively, correspond to the SINR-based, the *Relax*, the *Evolve* and the optimal solutions. Moreover, dashed-dotted, dashed and dotted lines correspond to low, medium and high backhaul capacity (i.e., C^{BH} equals 40/80/120 Mbit/s).

Note that in the first case, the backhaul is likely to be the main constraint on network performance (Figure 7.5(a)); hence, the classic SINR-based approach, which only takes into account the quality of the radio link, is characterized by poor performance. However, the higher the backhaul capacity, the lower its impact on the overall capacity: when C^{BH} is set equal to the maximum achievable RAN capacity

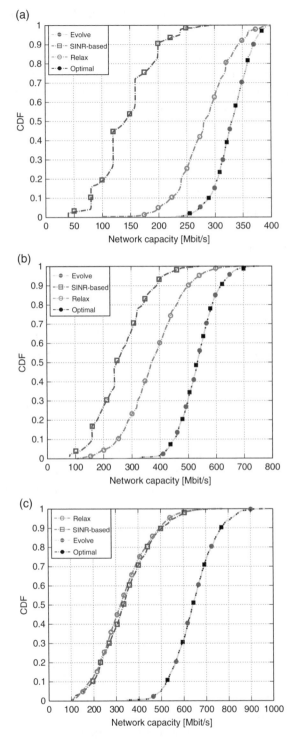

Figure 7.5 Cumulative distribution function of the network capacity achieved with different association schemes with respect to the backhaul capacity. (a) $C^{BH} = 40$ Mbps; (b) $C^{BH} = 80$ Mbps; (c) $C^{BH} = 120$ Mbps

($B \cdot \eta_{max}$ = 120 Mbit/s), only the quality of the radio links and the network load limit the performance. Therefore, the SINR-based approach achieves more valuable performance and it gains up to 133% with respect to the low-backhaul capacity case (compare plots using open squares in Figure 7.5(a) and Figure 7.5(c)). The *Relax* algorithm facilitates an improvement on the performance achieved by the classic SINR-based scheme in low-backhaul capacity scenarios, and our simulations show 97% of gain measured at the CDF median value (Figure 7.5(a)). However, increasing the backhaul capacity does not result in the *Relax* scheme showing further notable gains. This drawback is mainly due to two reasons: first, this approach is based on relaxing the constraint that forces each UE to be served by only one eNB; however, this may lead to a solution that diverges from the optimal one. Second, the *Relax* scheme does not guarantee to improve the network capacity during its iterative process. On the contrary, we note that *Evolve* has the same performance as the optimal solution (the filled circle and filled square plots are superimposed) and gains up to 132% and 100% with respect to the classic SINR-based and *Relax* schemes (measured at the median value of Figure 7.5(a) and Figure 7.5(c), respectively). The gain with respect to the SINR-based algorithm is due to the load and backhaul-aware properties of the proposed scheme that better balance service requests across the network and increase the overall resource utilization.

In the following, to investigate a more realistic scenario, we consider a dense small-cell deployment, where an MeNB, a cluster of SCeNBs and 30 UEs are located in each macro-cell sector (see Figure 7.6). Figure 7.7 shows a snapshot of the resulting association patterns by using the classic SINR-based scheme (Figure 7.7(a)) and the proposed *Evolve* algorithm (Figure 7.7(b)), respectively. In this example, *Cbk* is set equal to 40 Mbit/s and we have used an MCI scheduler. Simulation shows that *Evolve* increases the macro-cell offloading by moving part of the traffic from the highly loaded MeNBs to the surrounding lightly loaded SCeNBs. Furthermore, we can see that in Figure 7.7(a), 12/27 SCeNBs (and related backhaul facilities) are idle as they are not associated with any UEs; by contrast, in Figure 7.7(b), all the small cells are active, which results in higher resource utilization as well as improved network capacity.

Next we assess the performance achieved by the proposed *Evolve* algorithm with respect to the different resource-allocation policies and we evaluate the length of the iteration process with respect to different backhaul constraints. In Figure 7.8, results show the evolution of the average capacity in the three-sector macro-cell area. Filled-circle, square, and empty-circle marked lines, respectively, correspond to the MCI, RR and DRF policies. Moreover, solid, dashed and dotted-dashed lines, respectively, correspond to low, medium and high backhaul capacity. *Evolve* starts from the association found by the SINR-based algorithm (i.e., its first iteration) and iteratively improves the network performance towards the optimal solution, which results in (at least) 40% of gain in the investigated scenarios.

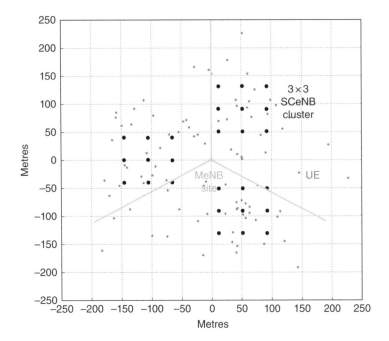

Figure 7.6   A snapshot of the 3×3 SCeNB clusters in the central macro cell

As expected, simulation results show that MCI policy enables higher network capacity with respect to both RR and DRF algorithms. In fact, MCI better exploits multi-user diversity by allocating resource to UEs characterized by better link quality. By contrast, in DRF policy, all UEs achieve the same data rate; hence, fairness is traded for the overall network performance.

Our simulations also show that the proposed *Evolve* algorithm converges, after a limited number of iterations, which results in reduced latency and computational costs. When the backhaul is characterized by low capacity, there are few possible solutions associated with an improvement in system ergodic capacity; hence, the algorithm reaches convergence in only 17 iterations. However, when the backhaul is characterized by higher capacity, the number of enhancing solutions increases, and *Evolve* converges in 32 iterations. Note that, in such a dense deployment scenario, we have measured that the *Relax* algorithm needs more than 100 iterations to converge.

Previous simulations have shown that the *Evolve* algorithm achieves the same performance as the optimal solution while limiting the overall latency and complexity. However, *Evolve* may prevent UEs from selecting the eNB associated with the best RSRP. Hence, we may expect that this approach will increase energy consumption due to uplink transmissions and reduce the mobile terminal lifetime.

Therefore, in the following, we aim to investigate the impact of the proposed *Evolve* algorithm to the uplink-radiated power. Fractional power control is used at

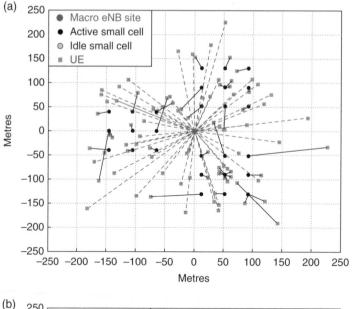

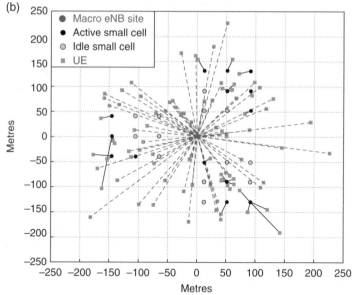

Figure 7.7   A snapshot of the association pattern when using as association metric the strength of the downlink signal (a) and the aggregated network load (b)

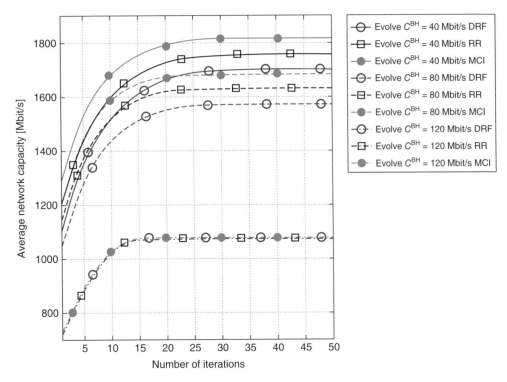

Figure 7.8 Convergence of the *Evolve* algorithm with respect to different backhaul constraints and resource-allocation policies

the UEs in the uplink transmissions [19] to mitigate inter-cell interference, and the emitted power can be evaluated as:

$$P_{UE} = \min\left(P_{max}, P_0 + 10 \cdot \log_{10} M + \lambda \cdot L\right)[dBm], \tag{7.18}$$

where P_{max} is the maximum transmission power (23 dBm), P_0 is a UE-specific parameter (−78 dBm), M is the number of assigned physical resource blocks (set here equal to 6) to the UE, λ is the cell-specific path-loss compensation factor (0.8) and L is the downlink path loss measured at the UE using the control channel. Hence, cell selection impacts the uplink power by changing the value of the L parameter, and the farther away the chosen eNB, the higher the power.

Figure 7.9 presents the CDF of P_{UE} associated with the *Evolve* (the plots indicated by filled circles) and the SINR-based (the plots indicated by open squares) approaches when using MCI policy with respect to different backhaul constraints. First, we can note that the actual consumed power is well below the maximum transmission power P_{max} (200 mW). In fact, dense small-cell deployment enables UEs located at the edge of the macro cell to connect to a nearby SCeNB instead of a faraway MeNB.

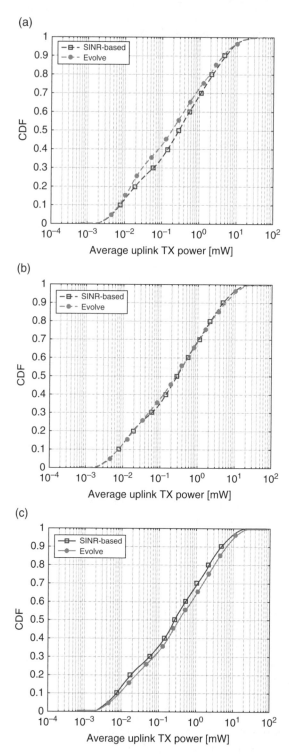

Figure 7.9 Cumulative distribution function of the uplink transmission power at UEs with different backhaul constraints. (a) $C^{BH} = 40$ Mbit/s MCI; (b) $C^{BH} = 80$ Mbit/s MCI; (c) $C^{BH} = 120$ Mbit/s MCI

Second, simulation results show that the investigated algorithms fairly require the same uplink-radiated power, and only a very limited percentage of UEs is characterized by a small increase in radiated power when using the proposed *Evolve* scheme (see Figure 7.9(c)). This increase in power is mainly due to the handover of UEs from nearby highly loaded SCeNBs to more distant lightly loaded SCeNBs. However, when the backhaul capacity is the main constraint (see Figure 7.9(a)), a high number of UEs hand over from the MeNB to closer SCeNBs, which reduces energy consumption.

7.5 Conclusion

Dense deployment of small cells is an enabling technology for handling high data rate demands required by future generations of wireless networks. In this scenario, heterogeneous backhaul solutions will be used to connect small cells with the core network, and access and backhaul have to be optimized jointly to enable efficient use of available resources. Up to now, cell selection has been based only on the quality of the radio link, which may limit the macro-cell offload and result in load congestion at cells with capacity-limited backhaul.

Accordingly, in this chapter, we have proposed to replace the classic SINR-based association criterion with an approach that jointly takes into account the cell load and the backhaul constraints. Therefore, we have analytically modelled the role of cell load, backhaul constraints and resource allocation in the cell-selection problem. Finally, we have presented two iterative algorithms, named *Evolve* and *Relax*, which associate UEs and eNBs according to the proposed approach. The *Evolve* algorithm has been proved to converge with a limited number of iterations to a near-optimal solution and simulation results show that it leads to notable system improvement with respect to the classical SINR-based algorithm, even in scenarios where the backhaul does not limit the network capacity. Further studies will extend our investigation by introducing cooperative transmissions and enabling each UE to be connected simultaneously to multiple eNBs.

References

[1] Hoymann, C., Larsson, D., Koorapaty, H. and Cheng, J.-F. (2013) A Lean Carrier for LTE. *IEEE Communications Magazine*, **51**(2), 74–80.

[2] Lopez-Perez, D., Güvenç, S., De La Roche, G., Kountouris, M. and Quek, T. Q. S. (2011) Enhanced intercell interference coordination challenges in heterogeneous networks. *IEEE Wireless Communications*, **18**(3), 22–30.

[3] Sun, S., Gao, Q., Peng, Y., Wang, Y. and Song, L. (2013) Interference management through CoMP in 3GPP LTE advanced networks. *IEEE Wireless Communications*, **20**(1), 59–66.

[4] 3GPP TSG RAN (2012) TR 36.927, 'Potential solutions for energy saving for E-UTRAN (Release 11),' v11.0.0, September.

[5] Next Generation Mobile Networks (NGMN) Alliance (2012) 'Small Cell Backhaul Requirements,' Backhaul Evolution, June.

[6] Rost, P., Bernardos, C. J., De Domenico, A., Di Girolamo, M., Lalam, M. *et al.* (2014) Cloud Technologies for Flexible 5G Radio Access Networks. *IEEE Communications Magazine*, **52**(5), 68–76.

[7] Parkvall, S., Dahlman, E., Ongren, G. J., Landstrom, S. and Lindbom, L. (2011) Heterogeneous network deployments in LTE. *Ericsson Review*, **2**.

[8] Madan, R., Borran, J., Sampath, A., Bhushan, N., Khandekar, A. and Ji, T. (2010) Cell Association and Interference Coordination in Heterogeneous LTE-A Cellular Networks. *IEEE Journal on Selected Areas in Communications*, **28**(9), 1479–1489.

[9] Guvenc, I. (2011) Capacity and Fairness Analysis of Heterogeneous Networks with Range Expansion and Interference Coordination. *IEEE Communications Letters*, **15**(10), 1084–1087.

[10] Lopez-Perez, D., Chu, X. and Guvenc, I. (2012) On the Expanded Region of Picocells in Heterogeneous Networks. *IEEE Journal on Selected Topics in Signal Processing*, **6**(3), 281–294.

[11] Ye, Q., Rong, B., Chen, Y., Al-Shalash, M., Caramanis, C. and Andrews, J. G. (2013) User Association for Load Balancing in Heterogeneous Cellular Networks. *IEEE Transactions on Wireless Communications*, **12**(6), 2706–2716.

[12] Olmos, J., Ferrus, R. and Galeana-Zapien, H. (2013) Analytical modeling and performance evaluation of cell selection algorithms for mobile networks with backhaul capacity constraints. *IEEE Transactions on Wireless Communications*, **12**(12), 6011–6023.

[13] 3GPP TSG RAN (2006) TR 25.814, 'Physical Layer Aspects for Evolved UTRA (Release 7),' v7.1.0, September.

[14] 3GPP TSG RAN (2013) TR 36.932, 'Scenarios and requirements for small cell enhancements for E-UTRA and E-UTRAN (Release 12),' V12.1.0, March.

[15] Pisinger, D. (1995) *Algorithms for knapsack problems*, PhD dissertation, University of Copenhagen.

[16] Feng, S. and Seidel, E. (2008) *Self-organizing networks (SON) in 3GPP long term evolution*. Nomor Research GmbH, white paper.

[17] Nohrborg, M. (n.d.) 'Self-Organizing Networks.' Available at: http://www.3gpp.org/technologies/keywords-acronyms/105-son.

[18] 3GPP TSG RAN (2010) TR 36.814, 'Evolved Universal Terrestrial Radio Access (E-UTRA); Further advancements for E-UTRA physical layer aspects (Release 9),' V9.0.0, March.

[19] Guvenc, I., Moo-Ryong, J., Demirdogen, I., Kecicioglu, B. and Watanabe, F. (2011) Range expansion and inter-cell interference coordination (ICIC) for picocell networks. In *Proceedings of the IEEE Vehicular Technology Conference*, San Francisco, pp. 1–6, September.

8

Multiband and Multichannel Aggregation for High-speed Wireless Backhaul: Challenges and Solutions

Xiaojing Huang
Faculty of Engineering and Information Technology, University of Technology Sydney (UTS), Australia

8.1   Introduction

In cellular and wireless broadband networks, backhaul is the communication link (wired and/or wireless) between a base station and the associated switching node. Sometimes a number of base stations can be connected to a switching node via a hub station, where each base station has at least one backhaul link to the hub station. Backhaul can also refer to other high-speed transmission links and networks which connect distributed sites and centralized points, such as a network backbone, enterprise connection, fibre extension, and so on.

Due to the ever-increasing capacity required to support high-speed broadband services, the backhaul network is under intensive pressure [1–3]. A number of challenges face such backhauls, the most significant one of which is how to achieve a higher data rate or capacity up to multiple Gigabits per second (Gbps). For example, if the capacity of a cell (or sector) in a broad wireless access base station is 1 Gbps, the backhaul capacity required by a three-sector base station would be at least 3 Gbps. Sometimes the traffic from multiple base stations will be aggregated before reaching

Backhauling/Fronthauling for Future Wireless Systems, First Edition.
Edited by Kazi Mohammed Saidul Huq and Jonathan Rodriguez.
© 2017 John Wiley & Sons, Ltd. Published 2017 by John Wiley & Sons, Ltd.

the core network. This will drive the backhaul capacity to a much higher rate, say 10–15 Gbps. The second challenge is the link distance of the backhaul. To deliver broadband services to unserved areas, such as rural and regional areas which are often quite remote from the main telecommunication infrastructure, a long-distance back-haul link is required. The third challenge is how to achieve low-latency communica-tions between end users across the backhaul networks. Although low latency has always been important for the delivery of high-quality voice, video and data services in broadband networks, recent application requirements within many industry sectors, such as gaming and finance, have brought low latency right to the forefront of the telecommunications industry.

Fibre is the primary medium to deliver leased synchronous digital services and Ethernet services. It is the first choice to offer high-data-rate backhauling from 155 Megabits per second (Mbps) to 10 Gbps capacity. However, due to high installation expenses, such as digging the trenches in which the fibre is to be laid, owning a fibre is a significantly more expensive capital expenditure (CAPEX) option. It is also esti-mated that leased lines currently account for roughly 15% of the network operating expenditure (OPEX). By contrast, wireless backhaul is more cost-effective than leased T1/E1, DS3 or OC-3 lines. In addition to the economic benefits of ownership, wireless backhaul also allows service providers to retain end-to-end control of their data and gain the security, stability and freedom associated with full control over their own net-work. For less populated rural areas, where the cost to lay fibre can be prohibitive, wireless backhaul will be the only viable solution. In addition, because radio propa-gates over the air faster than light travels through fibre, wireless backhaul can achieve lower latency than fibre. For backbone communication networks which provide com-munication links between space crafts or between space craft and a base station, wireless backhaul can propagate through clouds with long reach and high availability.

Therefore, if wireless backhaul can achieve multi-Gigabit data rates comparable to fibre capacity, it will be a cost-effective solution to fibre replacement in urban areas and a very attractive proposition for providing remote communities with broadband services [4]. It can also serve as, or be part of, an ultra-low-latency network for various low-latency applications.

However, there are some major technical challenges to developing a wireless back-haul link. The first challenge is how to achieve the required data rate. A wireless back-haul can operate at either microwave frequency bands or millimetre-wave (mmWave) frequency bands. At microwave frequencies, a radio frequency (RF) band typically has only about 200 MHz bandwidth in total and, given a location, many channels in the band may have been occupied by some existing services already. Since the bandwidth of each microwave channel is narrow (7 to 80 MHz), it is almost impossible to develop a multi-Gigabit link in microwave bands even with some smart channel bonding and aggregation techniques. A higher data rate is possible for a wireless link operating at mmWave frequency bands such as the E-band (71–76 GHz and 81–86 GHz) which has a contiguous bandwidth of 5 GHz in each lower and upper band.

Currently, commercial E-band wireless links can only provide up to 1.25 Gb/s data rates and employ low-order, two-state modulations such as amplitude shift keying (ASK) or binary phase shift keying (PSK) with low spectral efficiency. Using higher-order modulations to increase data rate and spectral efficiency for E-band systems has attracted significant interest in recent years from both the research community and industry. For example, a direct QPSK SiGe BiCMOS transceiver was reported in October 2010 by the University of Toronto; this can be used for realizing a nearly 10 Gbps E-band link over 5 GHz bandwidth [5]. Huawei announced its second-generation E-band backhaul system with a data rate up to 2.5 Gbps using 64-QAM over two 250 MHz channels in October 2012 [6].

Achieving both a high data rate and high spectral efficiency at the same time in the E-band requires the use of high-order modulation over the full 5 GHz spectrum, which is still a significant technical challenge. In terms of achieving high-order modulation, a digital modem should be implemented in programmable signal-processing devices such as ASIC or FPGA. However, digital processors and mixed signal devices which can cope with large bandwidth with sufficient performance are not available or are extremely expensive. Analogue circuitry can be applied to wider bandwidth, but the component tolerances, manufacturing fluctuations and other practical impairments limit the modulation to very low levels such as QPSK. In addition, there are some system penalties with high-order modulation, which include system complexity, design and production costs, receiver sensitivity reduction and output power reduction. The reduced transmit power and receiver sensitivity will also impact on the link distance.

The second challenge is the link distance. A microwave link can operate over a range of several tens of kilometres whereas a mmWave link can only typically reach several kilometres. The difference in operating range is due to the different radio propagation characteristics, such as atmospheric absorption and rain fade, in microwave and mmWave bands. Attenuation by atmospheric gases at specific radio frequencies depends on the atmospheric conditions such as barometric pressure, temperature, humidity and density of water droplets in clouds or fog. It is well known that with the exception of the 60 GHz band, where radio propagation is severely affected by atmospheric oxygen resonant absorption, specific attenuation increases proportionally to water vapour and droplet density. In the absence of precipitation, moderate specific attenuation at the E-band (below 3 dB/km) makes this band suitable for short- and medium-range wireless links. The main factor that limits available communication range at the upper microwave and mmWave frequencies is rain fade.

In this chapter, the various system architectures for multiband and multichannel aggregation, which are particularly useful for microwave backhaul systems, are discussed. A spectrally efficient channel-aggregation scheme is described for effective use of mmWave spectrum. The challenges for achieving high-speed wireless transmission and improving spectrum efficiency and power efficiency are addressed.

Various technical solutions are provided with detailed explanations and illustrations. Real-world systems using these techniques are introduced to demonstrate their practical applications.

8.2   Spectrum for Wireless Backhaul

Wireless communications use radio frequencies as the transmission medium. The radio spectrum is the range of all electromagnetic radiation frequencies. Traditionally, the radio spectrum in the microwave band covers the 6–40 GHz frequencies. Though the term mmWave refers to the radio frequency with a wavelength of less than 1 cm (or a frequency of 30 GHz and above), it is more convenient to refer to mmWave bands as those of 55 GHz and higher. This is because the microwave bands from 6–40 GHz are relatively consistent in characteristics and are managed in a similar way by regulators around the world [7].

8.2.1   Microwave Band and Channel Allocation

Microwave is the typical medium for wireless backhaul. The microwave spectrum is divided into specific bands administered by national regulators in individual countries. In the United States, the relevant government agency is the Federal Communications Commission (FCC). In Australia, it is the Australian Communications and Media Authority (ACMA). In Europe, the frequency bands are set up by the European Conference of Postal and Telecommunications Administrations (CEPT) and the technical rules for their use are defined by the European Telecommunications Standards Institute (ETSI). The bands are administered and managed by individual countries' national regulators.

A microwave band is normally divided into multiple channels, for which frequency division duplex (FDD) is used. Half of the channels are used for transmission and the other half, separated by a central guard band, for reception. The channel bandwidth is normally the same for all channels in a band, but may be different in different bands. A general radio frequency band plan is shown in Figure 8.1.

The 6 GHz and 11 GHz bands are widely used for wireless backhauling, but have very strict technical restrictions on their use. With a minimum antenna size of 2 m, the 6 GHz band is very suitable for long-range transmission, but the data rates are low due to narrower channel bandwidth (typically 29.65 and 40 GHz). The 18 and 23 GHz bands are widely used for higher-data-rate applications due to the larger channel sizes (up to 80 MHz) but with shorter range operation. The 28 and 38 GHz bands are managed very differently to the lower bands. They can also support high-data-rate transmission.

Compared with mmWave bands, the bandwidth of each microwave channel is narrow (7 to 80 MHz), and therefore the data rates of traditional microwave links are

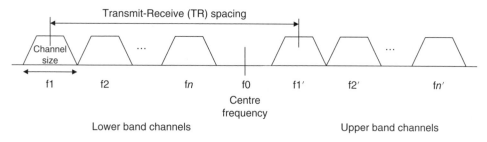

Figure 8.1 General microwave radio frequency band plan [7]

only up to several hundred Mbps. To achieve higher speeds, sufficient bandwidth is necessary. However, a large, contiguous bandwidth is not available at microwave frequencies for backhauling, except for some disjoint radio frequency bands and channels. A natural solution is to aggregate multiple channels to obtain the required bandwidth, as has been implemented in some vendors' products. However, such a straightforward combination of multiple low-rate wireless systems is neither cost-effective nor spectrally efficient. On the other hand, due to the radio propagation characteristics, the radio frequencies should be below 10 GHz for long-range operation. Achieving both high speeds and long range for wireless backhauls remains a significant technical challenge.

8.2.2 Millimetre-wave Band and Usage Trend

In the millimetre-wave bands, very wide channels are available with ample bandwidth to enable very high data rates for wireless systems. In the 60 GHz band, the FCC allocated 7 GHz of spectrum from 57 GHz to 64 GHz for unlicensed use. Other countries also have similar allocations in the 60 GHz band, with bandwidth varying from 5 GHz to 7 GHz. In the 70 GHz and higher millimetre-wave bands, there is a total of 13 GHz bandwidth available worldwide; that is, 71–76 GHz, 81–86 GHz and 92–95 GHz.

With wider bandwidth, the communication capacity can be increased in general. However, how to efficiently use the available bandwidth is still a challenge. Due to the limitations in digital-to-analogue (D/A) and analogue-to-digital (A/D) conversion speed as well as digital hardware resources, a wide bandwidth with multi-gigahertz spectrum may not be able to be accommodated in a single processing chain as one transmission channel. The multi-gigahertz spectrum may need to be divided into multiple smaller bandwidth channels, and hence channel aggregation is still necessary to make full use of the available bandwidth.

As 70/80 GHz E-band wireless systems become more and more popular for use in broadband networks with macro and small cells, it is expected that the E-band spectrum will be more and more congested. Unlike the unchannelized band plan in the

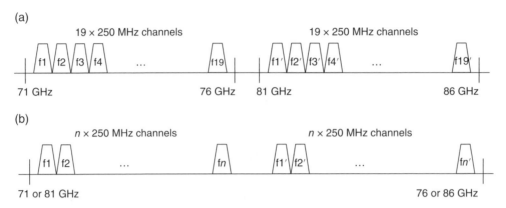

Figure 8.2 E-band channel plan for 10 GHz TR spacing (a) and less than 5 GHz TR spacing (b) [8]

US, each E-band in Europe has been divided into nineteen 250-MHz channels with a 125-MHz guard band at the two sides of each 5-GHz band in order to prevent possible interference from different E-band systems, as shown in Figure 8.2. Different channel combinations and arrangements for time domain duplexing (TDD) and frequency domain duplexing (FDD) applications are also defined [8]. Even smaller bandwidth channels such as 62.5 MHz have been recommended to allow more available links and more efficient spectrum usage. With such a frequency allocation plan and increasing demand for E-band links, the use of narrower bandwidth and higher-order modulation to offer multi-Gigabit data rates has become an industry trend for E-band system development. Therefore, E-band tends to be used in a similar way to conventional microwave bands.

8.3 Multiband and Multichannel Aggregation

This section discusses general architectures and implementations for multiband and multichannel aggregation. The basic assumption is that there are multiple frequency bands available for the backhaul system, each having multiple channels of narrow bandwidth. One band or a number of contiguous channels in a band can be sampled by A/D and D/A to form one digital data stream.

8.3.1 Band and Channel Aggregation Overview

To achieve high-data-rate wireless backhaul, a natural solution is to increase the transmission bandwidth by aggregating a number of narrowband channels in different frequency bands. In the wireless backhaul industry, it has been common practice to realize high-data-rate microwave relay systems by combining a number of low-rate wireless systems operating at different radio frequencies. For example, by simply stacking four systems, each having 150 Mbps capacity, up to a 600 Mbps data rate can be achieved. However, such a straightforward combination of multiple low-rate wireless systems has many drawbacks, as described below [9].

First, the direct combination of low-data-rate systems is not cost effective, as separate baseband processing modules, analogue-to-digital and digital-to-analogue converters and RF chains including mixers, bandpass filters and power amplifiers are required for individual RF channels. As a result, the system cost increases linearly with the data rate. To reduce the implementation cost, signals from multiple RF channels can be combined to form a multicarrier signal before power amplification. However, a multicarrier signal can exhibit a high peak-to-average power ratio (PAPR), which significantly reduces the power efficiency of the system since a transmit power back-off has to be enforced to reduce nonlinearity. In general, the higher the number of carriers used in the combined signal, the higher the PAPR will be.

Second, the direct combination of low-data-rate systems cannot make full use of the available bandwidth. As has been seen from the last section, a microwave frequency band is divided into a number of narrowband RF channels and the channels are arranged in pairs separated by a fixed transmit–receive duplex spacing. Co-channel and adjacent channel protection ratios are defined to prevent interference among different channels. If data are transmitted independently in different RF channels, guard bands must be enforced to prevent emission into adjacent channels, resulting in a loss of spectral efficiency.

The multiband and multichannel aggregation technique described in this chapter combines multiple RF channels (also called subbands hereafter) in multiple frequency bands to provide a high-data-rate wireless link with improved spectral efficiency, power efficiency and cost efficiency for broadband backhauling applications. The two novel ideas employed in this technique are subband aggregation and the use of single-carrier modulation in each aggregated subband (ASB). The subband aggregation merges multiple adjacent RF channels to form a wider bandwidth subband so that no guard band in each adjacent channel of the ASB is required and the spectral efficiency can be improved. For a given ASB, single-carrier modulation is applied to modulate data symbols on the centre frequency (or carrier) of the ASB. Since subband aggregation reduces the number of ASBs for a given frequency band, there are fewer modulated signal carriers to be combined to form a multicarrier RF signal and hence the peak-to-average power ratio (PAPR) of the RF signal to be amplified and transmitted in the given frequency bands can be reduced to achieve higher power efficiency.

Three alternatives for the multiband transceiver will be described. The first one directly uses time-domain single-carrier modulation and demodulation for each ASB in a frequency band. The second one uses frequency-domain digital baseband processing for a frequency band to generate and receive digital baseband signals, resulting in an orthogonal frequency-division multiplexing (OFDM) type system in which multiple subcarriers are used and the subcarriers within an ASB are precoded and decoded using the discrete Fourier transform (DFT). This system is called ASB-OFDM. The third one uses the software-defined radio approach which combines all the allocated RF channels in multiple bands in the digital domain, offering the ultimate capability and flexibility in aggregating radio spectral resources to provide high-data-rate wireless backhauling transmission.

8.3.2   System Architecture

The multiband, multichannel wireless link is mainly used for providing point-to-point communications between two base stations (BSs) or between a BS and a fixed access point (AP) within a broadband access network. For convenience, the BSs and APs are called nodes in the following. The wireless link between two nodes is illustrated as an example in Figure 8.3, where each node is equipped with a multiband ASB transceiver and a directional antenna. One node is connected to the Internet backbone and the other node is connected to a broadband access network. We call the transmission from the Internet backbone node to the remote node the 'forward path' and the transmission from the remote node to the Internet backbone node the 'return path'.

The multiband ASB transceiver at the node connected to the Internet backbone is shown in Figure 8.4. It is composed of a forward path transmitter which composes the data bits to form data packets and transmits the data packets to the remote node, a return path receiver which receives data packets as well as forward-path channel state information (CSI) from the remote node, and a diplexer which separates the signal path between the forward and return paths. The multiband ASB transceiver operates in full duplex mode with an FDD scheme, that is, it transmits and receives at the same time but in different frequency spectra. The multiband ASB transceiver at the remote node is the same as that connected to the Internet backbone, except that the operation frequencies for the transmitter and receiver are swapped.

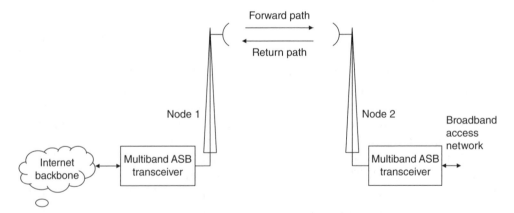

Figure 8.3   Point-to-point link using multiband ASB transceivers

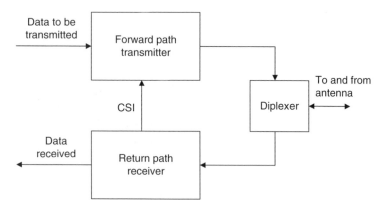

Figure 8.4   Multiband ASB transceiver architecture

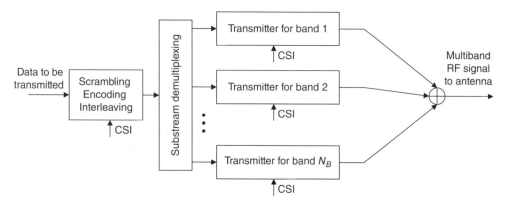

Figure 8.5   Forward path transmitter

The forward path transmitter, as shown in Figure 8.5, is composed of a scrambling, encoding and interleaving module, a substream demultiplexing module and N_B transmitters, each of which operates in a different frequency band. There is a total of N_B frequency bands used in the multiband ASB transceiver. Note that each band consists of a number of aggregated subbands. The input data bits are first scrambled, encoded using forward error codes (FECs) and interleaved to produce coded data bits. The coded data bits are then divided into N_B data substreams through substream demultiplexing. The coded data bits from each substream are mapped into data symbols and modulated on multiple frequency carriers or subcarriers to form an RF signal in a corresponding frequency band. Multiple RF signals formed in multiple frequency bands are finally combined and fed to the transmit antenna. The CSI received from the backward channel will be used to control each module in the multiband ASB transmitter to achieve adaptive modulation and coding (AMC) for optimized performance.

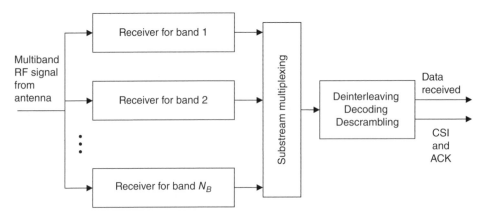

Figure 8.6   Return path receiver

The return path receiver is shown in Figure 8.6 and consists of N_B receivers, each of which operates in a different frequency band, a substream multiplexing module and a deinterleaving, decoding and descrambling module. The receiver operating in each frequency band receives the multiband RF signal and retrieves the coded data bits from data symbols modulated on multiple frequency carriers or subcarriers to recover each data substream. The substream multiplexing module then combines N_B recovered substreams to form a single coded data stream. The coded data stream is finally deinterleaved, decoded and descrambled to recover the original uncoded data bits. The CSI about the return path will be determined according to predetermined training sequences or obtained blindly. The CSI regarding the forward path is embedded in the received packet and will also be retrieved.

For a given frequency band, the data information needs to be modulated and up-converted from baseband to RF band, and the received RF signal needs to be down-converted and demodulated into baseband. There are two different architectures for signal up- and down-conversion. One is the in-phase/quadrature (I/Q) modulation architecture, by which the real and imaginary parts of the baseband data symbols are modulated to (or demodulated from) two orthogonal RF frequency carriers at the transmitter (or the receiver). The other is the digital IF architecture, by which the baseband modulation (or demodulation) is performed at digital baseband and a digital IF signal is converted to (or from) an analogue IF signal by a digital-to-analogue (or an analogue-to-digital) converter at the transmitter (or the receiver). The I/Q architecture only requires low-speed A/D and D/A devices, which is comparable to channel bandwidth, but it is usually subject to practical impairments such as I/Q imbalance. The digital IF architecture requires high-speed A/D and D/A devices without I/Q imbalance. The selection of an appropriate architecture should be made according to system performance requirements, implementation complexity and cost.

There are also two different signalling schemes for transmission and reception of data information in one signal processing chain, that is, single carrier and OFDM. The single-carrier scheme modulates data symbols directly onto a carrier frequency, whereas OFDM uses multiple subcarriers. Both have distinct pros and cons. The selection of an appropriate signalling scheme also needs to consider various aspects such as spectral efficiency, power efficiency, robustness against multipath propagation and channel fading, complexity and effectiveness of channel equalization, PAPR, out-of-band emission, sensitivity to sampling frequency offset and carrier frequency offset, and so on.

8.3.3 Subband Aggregation and Implementations

The novelty of the subband aggregation lies in the transmitter and receiver operating in a given frequency band in terms of how to use the multiple RF channels (subbands) to achieve higher spectral efficiency, power efficiency and cost efficiency.

As mentioned previously, a frequency band is divided into a number of RF channels in pairs according to regulatory rules. One RF channel in the pair is used for the forward path and the other is used for the return path. At any given site where the node is situated, all RF channels assigned to the same forward or return path are placed at either the lower or the higher block of the frequency band. For illustration purposes, Figure 8.7(a) shows all eight RF channels numbered 1 to 8 for the forward path used in a given frequency band, where only six channels, numbered 1, 3, 4, 6, 7 and 8 are assigned to the site, and channels 2 and 5 are assigned to a different licensee. To improve the spectral efficiency, the concept of subband aggregation is used, by which all neighbouring subbands are merged to form a wider subband called an aggregated subband (ASB). In the example shown in Figure 8.7(a), after subband aggregation, the number of subbands is reduced from six originally to three, where channels 3 and 4 as well as channels 6, 7 and 8 form two aggregated subbands respectively. A maximum allowed number of channels in an aggregated subband may be required to limit the bandwidth of an aggregated subband due to implementation difficulty and/or complexity issues in achieving a wide-band single-carrier modulation/demodulation, and there may exist different ways to aggregate subbands. This is illustrated in Figure 8.7(b) and (c), where channel 1 and channels 3 to 8 are assigned to the node and the maximum number of channels in an aggregated subband is assumed to be four. Channels 3 to 8 can be aggregated to form two wider subbands, each having three channels, or one having two channels and the other having the maximum four channels. To improve the power efficiency, single-carrier modulation for each aggregated subband is used, that is, the data symbols are modulated on the centre frequency (carrier) of each aggregated subband. All single-carrier modulated signals in a frequency band are combined to form the final RF signal to be power amplified and transmitted. Due to the subband aggregation and single-carrier modulation, there will

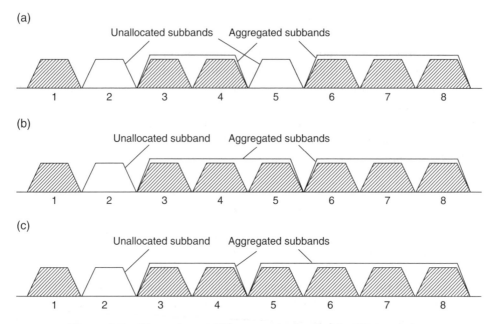

Figure 8.7   Illustration of different methods of subband aggregation

be fewer single-carrier modulated signals to be combined so that the PAPR of the final RF signal will be reduced as compared to that of the independently modulated and combined RF signal. In addition, the subband aggregation will also simplify the baseband signal processing and reduce the number of RF chains including the mixers, filters and amplifiers so that the implementation cost will be reduced.

Figure 8.8 illustrates the process of generating an RF signal by the transmitter for a given frequency band with single-carrier modulation for each aggregated subband and combining all modulated carriers. The aggregated subband arrangement shown in Figure 8.7(a) is used as an example for illustration. The carrier (i.e., the centre frequency) of each ASB is indicated in the ASB arrangement. The RF signal waveforms for the single-carrier modulated signals and the combined RF signal are also illustrated. With this signal-generation approach, the data symbols are directly modulated on the carrier of an ASB, and this is called aggregated subband frequency-division multiple access (ASB-FDMA). The block diagram of this transmitter is shown in Figure 8.9. The data bits from the substream are first distributed to each single-carrier modulation module. A single-carrier modulation module then modulates the data bits on to the carrier using digital modulation techniques such as multi-level phase shift keying (PSK) or quadrature amplitude modulation (QAM) to generate the RF signal to be transmitted in the corresponding ASB. A bandpass filter (BPF) follows in order to restrict the signal bandwidth to satisfy a desired transmit mask requirement. There are a total of N_s ASBs in the frequency band. Finally, all modulated RF signals to be

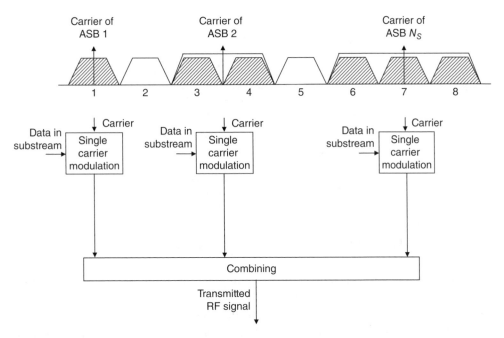

Figure 8.8   RF signal generation using a direct single-carrier modulation approach

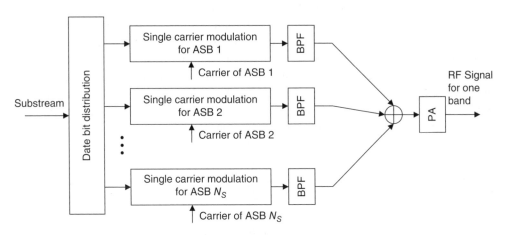

Figure 8.9   Transmitter block diagram using ASB-FDMA

transmitted in all ASBs are combined and amplified by a power amplifier (PA) to generate the RF signal for the frequency band.

The corresponding receiver to receive the signal generated by the transmitter using the ASB-FDMA approach is shown in Figure 8.10. It works as follows. The received multiband RF signal is first passed through a BPF to obtain the RF signal in the specified frequency band and then amplified by a low-noise amplifier (LNA). The amplified

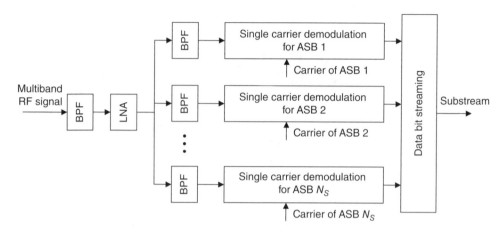

Figure 8.10   Receiver block diagram using ASB-FDMA

signal is then fed to each single-carrier demodulation module after being further filtered by a BPF to perform signal demodulation and equalization in the corresponding ASB. After demodulation and equalization, the received data bits in each ASB are obtained. Finally, all the received data bits are combined to form the received data substream.

Note that direct combining of RF signals may not be used in a particular implementation of the ASB-FDMA transmitter. An alternative approach is to combine the modulated carriers at some intermediate frequencies (IF) for each ASB and up-convert the combined IF signal to the right RF frequency band. Correspondingly, the receiver may first down-convert the received RF signal to an IF band and then demodulate the data symbols at respective IF carriers of the ASB. Also note that the signal carrier modulation and demodulation may use digital baseband processing and thus may include both digital domain and analogue domain modules.

The RF signal to be transmitted in a frequency band can also be generated using a frequency-domain multicarrier modulation approach. The process of this approach is illustrated in Figure 8.11, where the same aggregated subband arrangement shown in Figure 8.7(a) is used as an example. In this implementation, digital signal-processing techniques are used to generate a baseband signal first. For this purpose, the frequency band is shifted to baseband centred about the zero frequency (i.e., direct current) in the illustration (other implementations are possible, e.g., the second Nyquist zone). The entire frequency band is first divided into a number of subcarriers. The subcarriers which fall into an aggregated subband form a cluster of consecutive subcarriers. Data symbols to be transmitted are then allocated to the cluster of subcarriers and modulated using the single-carrier frequency-division multiple access (SC-FDMA) technique to generate a single-carrier baseband signal. All such single-carrier baseband signals generated by different clusters of subcarriers are combined to form the baseband signal for the frequency band. Finally, the baseband signal is shifted to

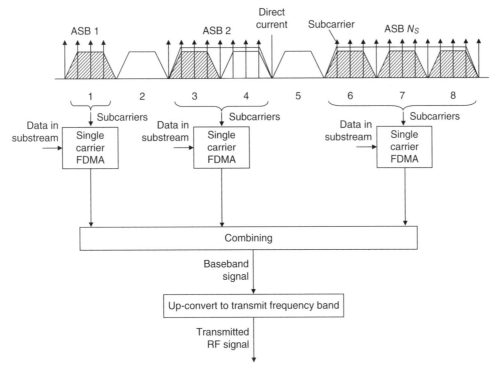

Figure 8.11 RF signal generation using the frequency-domain multicarrier modulation approach

the frequency band according to the centre frequency of the frequency band. In Figure 8.11, the baseband signal envelopes and the transmitted RF signal waveform are also illustrated. The block diagram of the transmitter which implements this frequency-domain multicarrier approach is shown in Figure 8.12, where the data bits from the substream are first allocated to different clusters of subcarriers and mapped into data symbols using symbol constellation-mapping techniques such as QAM. A set of data symbols corresponding to a cluster of subcarriers is then precoded using a discrete Fourier transform (DFT) to form a new set of precoded data symbols. After performing such precoding for all clusters of subcarriers, an inverse fast Fourier transform (IFFT) of size equal to the total number of subcarriers (including the null subcarriers in unallocated RF channels) is performed to produce the time-domain baseband signal. Note that the IFFT automatically realizes the combination of the multiple SC-FDMA signals, so a combining process shown in Figure 8.11 is inherent in the transmitter block diagram. After parallel-to-serial conversion (P/S) and cyclic prefix (CP) insertion or zero-padded (ZP) suffix appending, an OFDM-type symbol is generated. We call this signal-generation process ASB-OFDM. The ASB-OFDM symbol will be further converted into an analogue signal by a dual digital-to-analogue

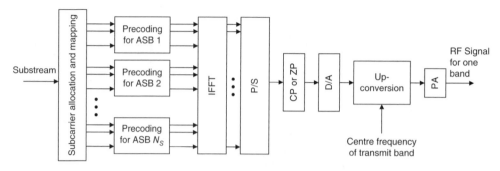

Figure 8.12 Transmitter block diagram using ASB-OFDM

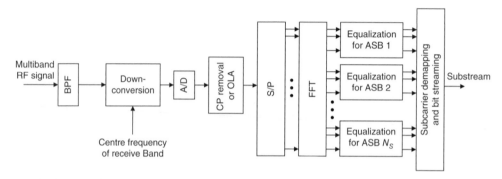

Figure 8.13 Receiver block diagram using ASB-OFDM

converter, shifted to the frequency band by an up-conversion module, and finally power amplified to generate the RF signal for the frequency band.

The corresponding receiver to receive the signal generated by the transmitter using ASB-OFDM is shown in Figure 8.13. Similar to the receiver using ASB-FDMA, the received multiband RF signal is first passed through a BPF to obtain the RF signal in the specified frequency band and then amplified by an LNA. However, the amplified signal is then shifted into baseband and converted into the digital domain by a dual analogue-to-digital converter (A/D) to obtain the received ASB-OFDM symbol. Before further processing, the CP of the received ASB-OFDM symbol will be removed (if CP is inserted in the ASB-OFDM symbol at the transmitter) or an overlap-add (OLA) operation will be performed to the received ASB-OFDM symbol (if ZP is appended to the ASB-OFDM symbol at the transmitter). The resulting ASB-OFDM symbol is then converted into the frequency domain after serial-to-parallel conversion (S/P) and fast Fourier transform (FFT). In the frequency domain, the subcarriers are grouped into clusters according to the aggregated subband arrangement in the frequency band. Each cluster of subcarriers is then equalized to compensate for the propagation channel effects and decoded using an inverse DFT (IDFT) matrix to recover the data symbols transmitted in the corresponding aggregated subband.

All the recovered data symbols are finally demapped into data bits and combined to form the received data substream.

Note that the up-conversion and down-conversion modules may use an appropriate IF stage to accommodate different implementation cost and complexity requirements.

The different implementations for the transmitter and receiver described above apply to all the frequency bands considered in the multiband ASB transceiver. However, the RF channel bandwidth and assignment as well as the aggregated subband arrangement can be different for different frequency bands. The system parameters, such as the total number of aggregated subbands, the total number of subcarriers and the subcarrier frequency spacing, can also be different for different frequency bands.

In all of the above implementations of a high-data-rate wireless transceiver, the subband aggregation can be dynamically performed, that is, given the information about the frequency band and subband assignment, the transceiver can automatically adjust the system parameters and/or reconfigure the hardware to achieve better performance. For example, when the transmitter buffer is full or nearly full, the data need to be transmitted at the highest data rate. In this case, all available subbands are aggregated and used for data transmission. When the transmitter buffer is nearly empty, there are fewer data to be transmitted and the data rate is low. In this case, fewer subbands can be aggregated and used for data transmission; the minimum number being one. In general, the number of aggregated subbands, the number of subbands in an aggregated subband and the number of frequency bands can be dynamically selected according to the data rate requirement. It may be the case that data will not be modulated into transmitted signals in all aggregated subbands at any given instant, again depending upon data rate requirements. Other system parameters such as the coding rate and modulation type can also be determined or adjusted according to different subband-aggregation schemes.

8.3.4 *Full SDR Approach for Band and Channel Aggregation*

The third implementation to realize a multiband, multichannel transmitter is illustrated in Figure 8.14, where the transmitter bandwidth covers all the N_B frequency bands to provide the ultimate capability and flexibility in aggregating radio spectral resources and offering high data transmission rates. In this implementation, the baseband is centred about the DC but the bandwidth is wide enough to accommodate the entire transmitted signal bandwidth over multiple frequency bands. The baseband is first divided into a number of subcarriers. The subcarrier spacing is fine enough to distinguish subbands (i.e., at least one subcarrier per subband). The subcarriers fall in unallocated subbands and/or bands are nulled. Data to be transmitted are scrambled, encoded, interleaved and then fed to an OFDM-type modulator to generate a modulated digital baseband signal. Only the subcarriers in the allocated subbands are modulated by the data symbols. The digital baseband signal is further up-converted to the

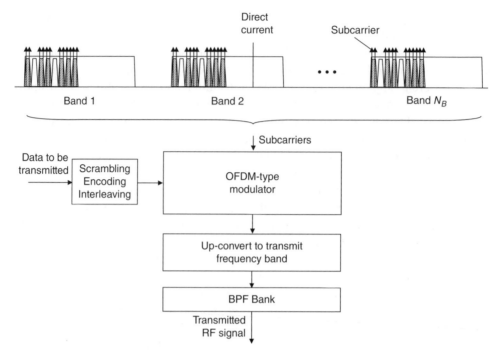

Figure 8.14   Full multiband, multichannel approach

transmit frequency band. To reduce interference in unallocated frequency bands and/ or subbands, a bandpass filter bank is used after frequency up-conversion. The filtered RF signal is finally power amplified and transmitted by a wide-bandwidth antenna.

The OFDM-type modulator has a structure similar to that shown in Figure 8.12 but with more subcarriers to cover a wider bandwidth. Encoded data are mapped on to the valid subcarriers in the allocated subbands. Techniques for PAPR reduction, such as the frequency-domain precoding mentioned in the second implementation, may also be applied. Modulated subcarriers including all nulled subcarriers are transformed to time-domain signal samples via IFFT. After P/S, CP insertion or ZP appending and D/A, the analogue baseband signal is formed.

The frequency up-conversion may use direct conversion with I/Q architecture for shifting the complex baseband signal to an RF signal. Alternatively, a real digital IF signal may be generated by the OFDM-type modulator and then up-converted to the transmit frequency band using a single mixer. The corresponding multiband, multi-channel receiver operates in a reverse direction, as described in Figure 8.14. The received RF signal is first filtered by a bandpass filter bank to obtain the signal in allocated subbands in all frequency bands, and then down-converted to baseband. The baseband signal is processed by an OFDM-type demodulator to recover the coded data bits. The coded data bits are then deinterleaved, decoded and descrambled to recover the original uncoded data bits.

The OFDM-type demodulator has a similar structure to that shown in Figure 8.13. After A/D, CP removal or OLA and S/P, FFT is performed on the received digital baseband samples to transform the baseband signal to the frequency domain. Equalization is then performed for subcarriers in all subbands, and finally the uncoded data bits are recovered.

The frequency down-conversion may also use direct conversion with I/Q architecture to shift the RF signal to a complex baseband signal, or it may use a single mixer to shift the RF signal to an IF signal which is then digitized for processing by the OFDM-type demodulator.

8.4 Spectrally Efficient Channel Aggregation

The methods discussed in the previous section are useful for achieving high-speed backhauling in microwave bands, where the division of channels in each band is pre-defined by regulatory authorities. Now, we consider the condition where a wide contiguous bandwidth is allocated for use but it has to be divided into a number of subchannels. For example, there is a contiguous 5 GHz bandwidth in the E-band (71–76 GHz and 81–86 GHz) but the current A/D and D/A speed is not fast enough to sample the whole bandwidth and the digital hardware is not powerful enough to process it as a single digital baseband. Spectrally efficient channel aggregation should be used in this case to achieve both high speed and high spectral efficiency for wireless transmission.

8.4.1 System Overview

Spectrally efficient channel aggregation is a novel frequency-domain multiplexing technique, by which the wide contiguous bandwidth is divided into several lower-bandwidth subchannels to accommodate the limitations of existing electronic devices, avoid some common practical impairments and ensure the real-time implementation of advanced digital signal-processing algorithms.

A simplified block diagram of the spectrally efficient channel aggregation scheme is shown in Figure 8.15. Each node at the two ends of the transmission link is composed of three major functional modules: a digital modem, an IF module and an RF front-end. Dual polarization can also be used to double the system throughput.

For the forward link transmitter, the digital modem receives Ethernet packets and generates digital IF signals in multiple digital channels with dual polarization. Each digital channel uses a multiple-giga-samples-per-second (Gsps) D/A converter and a bandpass filter to convert the digital IF signal to an analogue IF signal with narrower bandwidth for each polarization. For the return link receiver, the digital modem receives multiple channel analogue IF signals from the IF module and each polarization, and recovers the Ethernet packets. Each digital channel uses a multi-Gsps A/D converter to sample the analogue IF signal from each polarization.

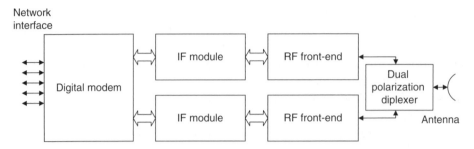

Figure 8.15   Block diagram of spectrally efficient channel aggregation system

The signalling scheme in the digital modem is selected as OFDM as it has inherent advantages such as higher spectral efficiency and simpler frequency-domain channel equalization. The major issue for OFDM is the larger PAPR, which will reduce the operation range. However, advanced PAPR-reduction techniques will be used in the proposed system to cope with this problem.

The IF module at the transmitter combines the IF signals generated by the digital modem for each polarization in multiple digital channels via frequency-domain multiplexing. The combined IF signal has the full bandwidth. At the receiver, the received full bandwidth IF signal is divided via frequency-domain demultiplexing into multiple IF channels for each polarization, each of which is then demodulated by the digital modem.

The RF front-end at the transmitter up-converts the IF signal to an RF signal centred at the E-band selected for the forward path and power amplifies the RF signal to sufficient power level. At the receiver, the RF front-end amplifies and down-converts the received RF signal centred at the E-band selected for the return path to an IF signal which is then fed to the IF module.

Other innovative technical approaches include frequency-domain multiplexing without a guard band, digital IF signal generation and reception and high-performance OFDM transmission, which are described in more detail in the next subsections.

8.4.2   Frequency-domain Multiplexing Without a Guard Band

When a contiguous bandwidth is divided into a number of smaller bandwidth subchannels, a guard band between adjacent subchannels is necessary to avoid inter-channel interference, since a sharp brick-wall filter at the output of each subchannel is impossible. This guard band will reduce the spectral efficiency significantly.

The novel frequency-domain multiplexing technique without a guard band combines signal processing in the digital domain and overlapped filtering in the analogue domain. As a result, the contiguous bandwidth can be fully utilized and high spectral efficiency is achieved. Figure 8.16(a) and (b) illustrate the frequency-domain multiplexing and demultiplexing process, where N denotes the number of subchannels and

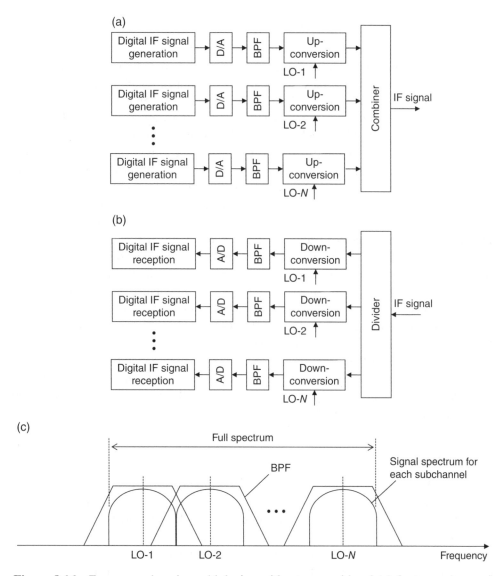

Figure 8.16 Frequency-domain multiplexing without a guard band (a) for transmitter and (b) demultiplexing for the receiver; (c) the full spectrum at IF

LO-1, LO-2 and LO-N denote the local oscillator (LO) frequencies for respective subchannels. Digital IF signals will be generated and received in the digital domain for a subchannel, whereas the analogue bandpass filter (BPF) will have wider bandwidth. By up-conversion or down-conversion of the IF signals with different LOs for respective subchannels, the full bandwidth signals can be combined or divided into individual subchannels without a guard band.

8.4.3   Digital IF Signal Generation and Reception

When high-order digital modulation is used, conventional implementation requires that each subchannel provides a digital I/Q baseband signal. After D/A and lowpass filtering (LPF), the analogue baseband signal is modulated by an I/Q modulator onto a carrier frequency. The novel digital IF signal-generation and reception technique uses a different approach which is advantageous in two ways as compared with the conventional approach. First, each subchannel provides a modulated digital IF signal at a suitable IF frequency rather than a digital baseband signal (i.e., centred at DC). Second, after D/A and BPF, the analogue IF signal is up-converted, rather than I/Q modulated, to a respective LO frequency for frequency-domain multiplexing. The advantages are as follows. First, only one D/A device is required for each channel after using a digital IF signal instead of a digital I/Q baseband signal, by which two D/A devices are required for each channel. Since the signal spectrum is very sharp with digital signal processing, the BPF specification is also relaxed. Second, digital modulation does not have the I/Q imbalance problem which is a significant impairment for the analogue I/Q modulator. Even if the I/Q imbalance can be compensated at the demodulator, it will increase the processing complexity and the compensation performance depends on how accurately the I/Q imbalance is modelled. Therefore, from the viewpoints of both complexity and performance, using a digital IF signal in the system is a more effective approach to realizing a high-capacity link.

A block diagram illustrating the digital IF signal generation and reception for each subchannel is provided in Figure 8.17, where an OFDM scheme is used as the fundamental signal-transmission technique.

8.4.4   High-performance OFDM Transmission

OFDM is a widely used multicarrier radio communication technique which offers high spectral efficiency, robustness against multipath propagation and channel fading and efficient frequency-domain channel equalization. It is adopted in this spectrally efficient channel aggregation to achieve the required spectral efficiency because cross-polarization interference cancellation (XPIC) necessary for dual polarization can be more efficiently realized in the frequency domain. However, there are some major disadvantages associated with OFDM transmission, such as PAPR, significant out-of-band emission (OOBE) and sensitivity to sampling frequency offset (SFO), carrier frequency offset (CFO) and phase noise.

A number of techniques have been proposed in the literature to overcome the above-mentioned disadvantages and improve OFDM transmission performance. For example, to reduce PAPR, various techniques, such as clipping, coding, phase optimization, nonlinear companding, tone reservation and tone injection, constellation shaping, partial transmission sequence and selective mapping, can be used. To reduce

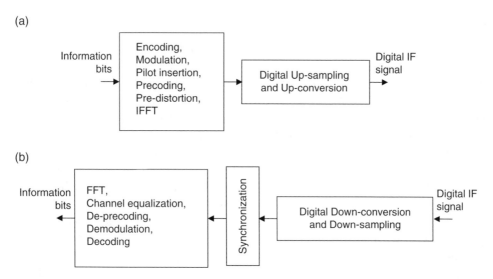

Figure 8.17 Digital IF signal generated at (a) transmitter and received at (b) receiver based on OFDM transmission

OOBE, various techniques, such as notch filtering, guard band reserving, time-domain windowing and cancellation with dedicated subcarriers, have also been proposed. Compensation for SFO, CFO and phase noise has also been extensively investigated and numerous methods have been proposed.

The above-mentioned techniques are mostly used independently to tackle one or more aspects of OFDM's drawbacks. Some of them may have conflicting effects. For example, the clipping method for reducing PAPR introduces both in-band distortion and out-of-band radiation, which increase OOBE. Notch filtering can reduce OOBE but it can also cause peak re-growth leading to higher PAPR.

Comprehensive solutions to these problems can be implemented in spectrally efficient channel aggregation. These solutions make possible a high-performance precoded OFDM system, where a block of data symbols is precoded using a precoding matrix, and the precoded output is then assigned to subcarriers for further processing. More details can be found in [10, 11].

8.5 Practical System Examples

The principles of achieving high-speed backhauling by multiband and multichannel aggregation have been described in the above sections. This section introduces some real-world systems developed at the Commonwealth Scientific and Industrial Research Organisation (CSIRO), in which these aggregation techniques are applied in practice.

8.5.1 CSIRO Ngara Backhaul

The first example is the CSIRO Ngara backhaul [9]. It operates in microwave frequency bands and achieves both high speed (up to 10 Gbps) and long range (over 50 km). The requirement for long range demands that the radio frequencies should be below 10 GHz. To achieve high speed, sufficient bandwidth is necessary. However, a large contiguous bandwidth is not available at microwave frequencies for backhauling, except for some disjoint RF bands and channels. A natural solution is to combine multiple single-channel radios to obtain the required bandwidth. However, such a straightforward combination of multiple low-rate wireless systems has many drawbacks, such as high cost, high complexity and low spectral efficiency. In order to solve these problems, CSIRO Ngara backhaul uses channel aggregation to merge multiple adjacent RF channels to form a wider bandwidth subband called an aggregated subband (ASB), as mentioned previously, so that guard bands within an ASB are not required and the spectral efficiency can be improved. For a given ASB, SC-FDMA is applied to modulate data symbols on the subcarriers of the ASB. The channel aggregation also reduces the number of ASBs for a given frequency band, so there are fewer modulated single-carrier signals to be combined to form a multicarrier RF signal and hence the PAPR of the RF signal to be amplified in the given frequency bands can be reduced to achieve higher power efficiency and, hence, longer range.

A system block diagram of the multiband transceiver is shown in Figure 8.18, which consists of a network interface, a digital subsystem and an RF subsystem.

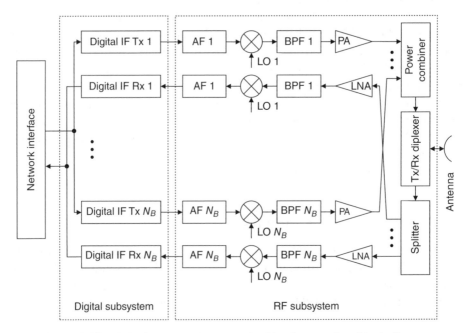

Figure 8.18 CSIRO Ngara microwave backhaul transceiver block diagram

The digital subsystem generates and receives digital IF signals in different frequency bands. The RF subsystem converts IF signals to/from RF bands and amplifies and transmits/receives RF signals. In each frequency band, analogue anti-aliasing filters (AFs) are used to eliminate the unwanted Nyquist responses and image responses. Bandpass filters (BPFs) are used to select signals from the frequency band after/before the up/down conversion with an appropriate LO frequency.

Based on the radio propagation characteristics, the microwave frequency bands at 6, 6.7 and 8 GHz have been selected for the implementation of Ngara microwave backhaul. When all the RF channels in the three bands are aggregated, the total bandwidth is over 724.75 MHz. High-order modulation such as 256-QAM is used to achieve a spectral efficiency of nearly 14 bits/s/Hz with a data rate up to 10 Gbps. The system uses a 3.6 m antenna and a transmit power of 26 dBm that is easily achievable from a commercial power amplifier with adequate linearity. An antenna height of 100 m above flat ground on average terrain is sufficient to eliminate diffraction fading at the range of about 60 km. For a 100-km range, antennas need to be elevated to 250 m using microwave towers combined with terrain topography.

8.5.2   CSIRO High-speed E-band Systems

Other examples include various high-speed E-band systems developed at CSIRO over the past ten years, such as the 6 Gbps system [12, 13], the 10 Gbps low-latency system and the 50 Gbps system [14].

The 6 Gbps system was first demonstrated in 2006. A simplified block diagram of this system is shown in Figure 8.19. The system includes a network interface, a digital modem including multiplexing and demultiplexing, an intermediate frequency (IF) module and a wideband mmWave front-end with a high-directivity antenna. The transmit and receive signals are combined using either a frequency- or time-domain diplexer. At the transmitter (Tx) input, the digital data stream is demultiplexed into N digital channels (e.g., $N=4$ to 16). Each digital channel is generated by a field-programmable gate array (FPGA) and a high-speed D/A. At the modulator, several identical high-data-rate digital channels are generated by direct computation of the analogue frequency signals. Analogue IF signals of each digital channel are multiplexed in the frequency domain in a spectrally efficient manner without a need for frequency guard bands between the adjacent channels. The combined IF signal is up-converted into a mmWave carrier frequency amplified by a power amplifier (PA) and transmitted over a line-of-sight (LOS) path. At the receiver (Rx), the received signal is passed through a bandbass filter (BPF) and a low-noise amplifier (LNA) and down-converted from the mmWave carrier frequency into IF and demultiplexed in the frequency domain into multiple channels, then sampled by the high-speed A/D converters and decoded by the FPGA into digital channels and then multiplexed into a single digital stream.

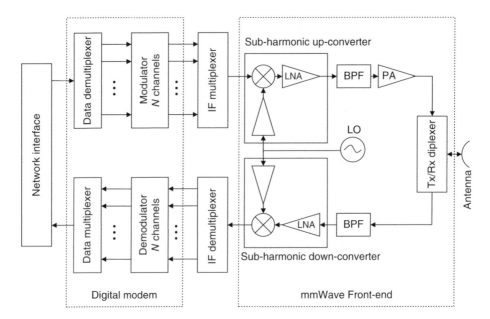

Figure 8.19   CSIRO 6 Gbps E-band backhaul transceiver block diagram

The mmWave transceiver uses heterodyne architectures with sub-harmonic frequency translation. The implementation of the sub-harmonic local oscillator (LO) allows for a reduction in complexity and cost of the transceiver. While a sub-harmonic mixing incurs a small penalty of several decibels in conversion gain or dynamic range, it provides a benefit of inherent suppression of both fundamental and even harmonics of the LO and down-converted LO noise.

The communication range for the CSIRO 6 Gbps system exceeds 5 km with output power of 17 dBm available using a single commercial MMIC amplifier. The range can be increased using the transmit power level up to the regulatory limit of 33 dBm.

The low-latency E-band system provides extremely low processing delay at a relaying transceiver and long range per hop so that the end-to-end latency for a multiple-hop wireless link can be shorter than an optical link while maintaining similar performance to that provided by an optical link. This is achieved by reducing the digital signal-processing chain and improving the power efficiency. The system adopts an I/Q modulation architecture at digital baseband with single-carrier transmission over the entire 5 GHz bandwidth. No frequency-domain multiplexing or digital IF is necessary. However, due to the availability of wide-bandwidth I/Q frequency converters at the IF stage, only up to 4.25 GHz bandwidth is actually utilized with a symbol rate of only 3.75 giga symbols per second, which limits the data rate to 10 Gbps.

The 50 Gbps E-band system has very similar system architecture to that used in the CSIRO 6 Gbps system which uses frequency-domain multiplexing and digital IF. However, it is different in the following aspects. First, for each digital channel, single-carrier transmission with root-raised-cosine pulse shape is used in the CSIRO 6 Gbps system. With 0.25 roll-off factor, a 625 MHz channel can only transmit data at a symbol rate of 500 Mbps. Though multiple 625 MHz channels are multiplexed at IF without a guard band, the available spectrum is still not fully utilized. In the 50 Gbps system, OFDM transmission is used in each subchannel and all subchannels are made orthogonal so that all available bandwidth is fully utilized. For example, for a 1.25 GHz subchannel, the symbol rate will be 1.25 giga symbols per second without sacrificing any spectral efficiency. Second, with advances in FPGA and data-conversion technologies, the 50 Gbps system uses higher-capacity FPGA and higher-speed A/D and D/A devices, enabling a wider bandwidth for each subchannel and a smaller total number of subchannels in the full 5 GHz bandwidth. This helps in reducing the PAPR of the RF signal to be transmitted and achieving higher power efficiency (and thus longer range). It also reduces the system complexity significantly. The RF front-end architecture is very similar to that utilized in the CSIRO 6 Gbps system with the exception of the LO source. In the proposed system, the RF front-end includes a 4x LO frequency multiplier. A reduction in the LO source frequency from 1/2 to 1/8 of the fundamental LO frequency allows for further reductions in complexity and cost of the LO circuitry.

Because the bandwidth for each subchannel is wider, practical impairments such as phase noise and jitter prevent the use of modulation above 64-QAM, so that the spectral efficiency cannot be as high as that for the Ngara microwave backhaul. However, the power efficiency of the 50 Gbps system is higher since the number of subchannels is much lower than that of the ASBs in the Ngara microwave backhaul. In addition, the complexity of the digital modem in the 50 Gbps system is simpler than that in the Ngara microwave backhaul since the MAC layer link aggregation will be easier and baseband signal sample rate conversion will no longer be necessary.

Compared with the low-latency system, the 50 Gbps system has significantly higher spectral efficiency and data rate. Also, due to the use of digital IF, the system does not suffer from the practical impairments such as I/Q imbalance inherent with the I/Q modulation architecture. Another advantage of digital IF is the elimination of the baseband stage impairments related to the bandwidth of differential to single-ended signal converters required at the interface between the A/D (D/A) device and the IF stage. By employing OFDM transmission and digital IF, cross-polarization interference cancellation (XPIC) is much easier to implement than using I/Q modulation.

On the other hand, the 50 Gbps E-band system does have the disadvantage of lower power efficiency due to the higher-order modulation and combination of multiple subchannels. More power back-off at the transmit power amplifier is necessary due to linearity requirements. This disadvantage can be partly mitigated by advanced PAPR-reduction techniques. In the long run, research efforts are needed for the development of nonlinearity compensation techniques, high power amplification devices and beamforming techniques using antenna arrays.

8.6 Conclusions

With the advance of broadband wireless access and next generation mobile systems, huge demands are being placed on backhaul infrastructure. As cost-effective alternatives to fibre backhaul, high-speed and long-distance wireless backhauls are becoming increasingly attractive. In addition to the aggregation of multiple microwave bands and channels to achieve high-speed transmission, the large contiguous bandwidth available in the millimetre-wave bands also allows for multi-gigabit data rate wireless backhaul applications. The various techniques presented in this chapter provide effective solutions to meet the backhaul requirements for the next generation of wireless broadband networks.

References

[1] Little, S. (2009) Is Microwave Backhaul Up to the 4G task? *IEEE Microwave Magazine*, **10**(5), 67–74.

[2] Parkvall, S., Furuskar, A. and Dahlman, E. (2011) Evolution of LTE toward IMTAdvanced. *IEEE Communications Magazine*, **49**(2), 84–91.

[3] Chen, S. and Zhao, J. (2014) The requirements, challenges, and technologies for 5G of terrestrial mobile telecommunication. *IEEE Communications Magazine*, **52**(5), 36–43.

[4] Wells, J. (2009) Faster Than Fiber: the Future of Multi-Gb/s Wireless. *IEEE Microwave Magazine*, **10**(3), 104–112.

[5] Sarkas, I., Nicolson, S. T., Tomkins, A., Laskin, E., Chevalier, P., Sautreuil, B. and Voinigescu, S. P. (2010) An 18-Gb/s, Direct QPSK Modulation SiGe BiCMOS Transceiver for Last Mile Links in the 70–80 GHz Band. *IEEE Journal of Solid-State Circuits*, **45**(10), 1968–1980.

[6] Huawei (2012) 'Huawei Debuts 2nd-Generation Ultra-Broadband EBand Microwave.' Press release, 2 October 2012. Available at: http://www.huawei.com/en/about-huawei/newsroom/press-release/hw-194598-e-bandmicrowave.htm

[7] Wells, J. (2010) *Multi-Gigabit Microwave and Millimeter-Wave Wireless Communications*, Artech House.

[8] ITU-R (2012) 'Radio-Frequency Channel and Block Arrangements for Fixed Wireless Systems Operating in the 71–76 and 81–86 GHz Bands.' Recommendation ITU-R F.2006, March.

[9] Huang, X., Guo, Y. J., Zhang, J. and Dyadyuk, V. (2012) A Multi-Gigabit Microwave Backhaul. *IEEE Communications Magazine*, **50**(3), 122–129.

[10] Huang, X., Zhang, J. and Guo, Y. J., (2014) Comprehensive Imperfection Mitigation for Precoded OFDM Systems. Paper presented at *IEEE 2014 International Conference on Communications (ICC2014)*, Sydney, Australia, 10–14 June.

[11] Huang, X., Zhang, J. and Guo, Y. J., (2015) Out-of-Band Emission Reduction and Its Unified Framework for Precoded OFDM. *IEEE Communications Magazine*, **53**(6), 151–159.

[12] Dyadyuk, V., Bunton, J. and Guo, Y. J., (2009) Study on High Rate Long Range Wireless Communications in the 71–76 and 81–86 GHz Bands. Paper presented at the *39th Europe Microwave Conference*, Rome, Italy, 28 September–2 October.

[13] Dyadyuk, V., Bunton, J., Pathikulangara, J., Kendall, R., Sevimli, O. *et al.* (2007) A Multi-Gigabit mmWave Communication System with Improved Spectral Efficiency. *IEEE Transactions on Microwave Theory and Techniques*, **55**(12) 2813–2821.

[14] Huang, H., Guo, Y. J., and Zhang, J. (2014) Multi-Gigabit Microwave and Millimeter-Wave Communications Research at CSIRO. Paper presented at the *14th International Symposium on Communications and Information Technologies (ISCIT2014)*, Incheon, Korea, 24–26 September.

9

Security Challenges for Cloud Radio Access Networks

Victor Sucasas, Georgios Mantas and Jonathan Rodriguez

Instituto de Telecomunicações, Aveiro, Portugal

9.1 Introduction

The recent explosive growth in mobile data traffic, the continuously growing demand for higher data rates and the steadily increasing pressure for higher mobility have led to the new generation of mobile communications, which is referred to as 5G. Over the past few years, a lot of effort has been put into developing this new generation of mobile communications with the vision of it being deployed by 2020 [1, 2]. 5G mobile communications aim to achieve big data bandwidth, infinite capability of networking and extensive signal coverage to support a plethora of high-quality personalized services to subscribers, while at the same time reducing the capital and operating expenditures (i.e., CAPEX and OPEX) of mobile operators. With a view to this aim, 5G communication systems will integrate a wide spectrum of technologies. Several of them are already available today and many others will be developed and deployed over the coming years [1–3].

The cloud radio access network (C-RAN) is a novel mobile network technology that has emerged as a promising solution to boost network capacity, enable energy-efficient operation, improve mobile network coverage and to reduce operating and capital costs in future mobile communication systems. In particular, C-RAN can increase network capacity by performing load balancing and cooperative processing of signals originating from multiple base stations (BSs) [4]. In addition, C-RAN can reduce power consumption by allowing BSs to be turned to low power or even shut

Backhauling/Fronthauling for Future Wireless Systems, First Edition.
Edited by Kazi Mohammed Saidul Huq and Jonathan Rodriguez.

down selectively. This is possible since all the baseband-processing tasks of local BSs can be migrated to a remote central entity, called a BBU pool, where many virtual baseband units (BBUs) are hosted and provide all the required processing functionality. In addition, the BBU pool allows operators to cover more service areas by only installing new RRUs connected with it. Moreover, the fact that the baseband-processing tasks are not allocated to RRUs but to the remote central BBU pool saves a lot in terms of operation and management costs. Furthermore, C-RAN shares equipment (e.g., transmission devices and GPS) more effectively, and thus capital expenditure is reduced as well. Therefore, C-RAN technology is expected to be a key ingredient in upcoming 5G communication systems [2, 4, 5].

However, despite its tremendous advantages, C-RAN technology has to deal with many inherent security challenges associated with virtual systems and cloud computing technology, which may hinder its successful establishment in the market [6]. In particular, it is worthwhile mentioning that the centralized architecture of C-RAN technology suffers from the single point of failure issue. Therefore, it is critical to address these challenges to enable C-RAN technology to reach its full potential and foster the deployment of future 5G mobile communication systems. In this sense, this chapter presents representative examples of potential threats and attacks against the main components of the C-RAN architecture in order to shed light on the security challenges of C-RAN technology and provide guidance that will ensure security in this emerging technology.

Following the introduction, this chapter is organized as follows. In Section 9.2, we give an overview of the C-RAN architecture based on the current related work on C-RAN technology; Section 9.3 provides representative examples of potential intrusion attacks in a C-RAN environment; Section 9.4 discusses examples of possible distributed denial of service (DDoS) attacks against the C-RAN architecture; and finally, Section 9.5 summarizes and concludes this chapter.

9.2   Overview of C-RAN Architecture

The general architecture of C-RAN, as depicted in Figure 9.1, consists of the following three main components [5]: (i) the remote radio units (RRUs) co-located with the antennas, (ii) the centralized BBU pool and (iii) the fronthaul network which connects the RRUs to the BBU pool [4, 5, 7].

The RRUs transmit the RF signals to user equipment (UEs) (e.g., smartphones, tablets) in the downlink or forward the baseband signals from UEs to the centralized BBU pool for further processing in the uplink. Mainly, the RRUs perform analogue-to-digital conversion, digital-to-analogue conversion, up/down conversion, filtering, RF amplification and interface adaptation [5]. On the other hand, the BBU pool consists of a number of BBUs which operate as virtual base stations to process baseband signals and optimize the network resource allocation for one RRU or a set of RRUs. The BBU assignment for each RRU can be implemented in a distributed or

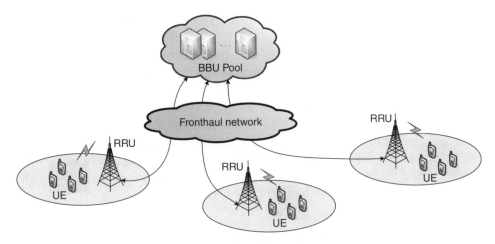

Figure 9.1   C-RAN architecture

centralized way based on the different resource management applied in the BBU pool. According to the distributed approach, one RRU directly connects to its corresponding BBU. Although the distributed approach is easier to implement, it does not reap the benefits of joint signal processing and central control. By contrast, according to the centralized approach, all RRUs connect to a central entity (i.e., the BBU pool). This approach provides many advantages in terms of flexible resource sharing and energy efficiency. Moreover, the centralized processing enables the implementation of efficient interference avoidance and cancellation algorithms across multiple cells [5]. Finally, the fronthaul network spans from the remote RRUs to the BBU pool and enables the C-RAN architecture. The fronthaul network is responsible for transporting the unprocessed RF signal from the remote antennas to the virtual BBUs. Although the fronthaul network requires a higher bandwidth, lower latency and more accurate synchronization than backhaul, connecting the BBU pool with the mobile core network, it enables a more efficient use of RAN resources. The fronthaul links can be made of different technologies such as fibre and wireless. However, optical fibre is considered the ideal means of fronthaul (i.e., without bandwidth constraints) for C-RAN, since it can provide large bandwidth and high data rates. For example, the NG-PON2 has a bandwidth of 40 GHz and 10 Gbps for the downstream and upstream respectively, and a range up to 40 km [4, 5].

9.3   Intrusion Attacks in the C-RAN Environment

The cloud RAN environment is, at its core, virtualized and, as a result, it is vulnerable to similar threats as virtual systems. In a virtual system, a malicious entity that breaks into the virtual environment can monitor, eavesdrop, modify or run software routines in multiple ways and with different purposes while remaining undetected. The C-RAN environment is not exempt from intrusion attacks, which may be even more harmful than in common virtual systems since C-RAN is the common control point for numerous mobile users and, hence, manages a tremendous quantity of resources and data. This is indeed motivation for malicious entities to attack C-RANs, since breaking into the C-RAN environment could give access to a plentiful set of personal data and virtualized resources.

In this section, we describe potential C-RAN intrusion vulnerabilities, depicting the possible causes that could lead malicious entities to take control of part of the C-RAN or run malicious routines inside its virtual environment. We also describe the technical difficulties for detecting intruders and limiting the scope of their actions in virtual environments. Finally, we detail the effects of intrusion attacks in order to elucidate the motivation of malicious entities to invade C-RANs.

9.3.1   Entry Points for Intrusion Attacks

A malicious entity can get into the C-RAN environment by exploiting vulnerabilities of the network infrastructure hosting the virtual environment. Specifically, an intruder can exploit: software vulnerabilities from patches and applications; misconfigured introspection and hypervision mechanisms; and rogue rollback procedures. It is also possible, although less likely, that an attacker could physically access C-RAN infrastructure which is installed outdoors [7]. An attacker that has successfully accessed the virtual environment and taken control of a virtual instance inside the host system can install a back door to spread malware inside the C-RAN, enabling a diverse set of intrusion attacks.

Once an attacker has gained access to the C-RAN, the attacker has the potential to eavesdrop, modify or run software routines. An intruder can initiate rogue virtual systems, such as virtual routers, databases and virtual BBUs, which would be seen as licit instances in the C-RAN virtual environment, hence enabling the attacker to misbehave undetected. It is worth noticing that the capability of the C-RAN environment to initiate on-demand virtual resources strengthens the intrusion attack strategies, since the attacker can get a malicious system inside the environment simply by executing a software routine, whereas in a conventional RAN architecture it would require the installation of malicious hardware inside the network infrastructure itself.

Figure 9.2 shows the vulnerabilities that intruders can use to get inside the C-RAN environment and the entities that can be controlled or subverted by a malicious entity:

- Virtual routers;
- Software routines;
- Virtual BBUs;
- Databases;
- The network infrastructure;
- The configuration and settings files;
- Access policies and firewall settings.

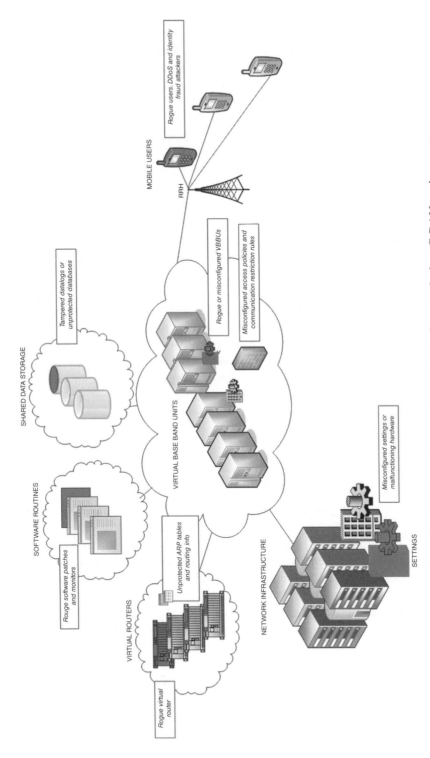

Figure 9.2 Vulnerabilities and entry points for intrusion attacks in a C-RAN environment

9.3.1.1 Software Vulnerability Exploitation

C-RAN is subject to the software vulnerabilities inherited from the software entropy, which is implicit in any highly complex software architecture. Software entropy is an unavoidable result of the technical debt [8] inherent in any software design, and hence it is present in any software-based system. C-RAN, as a software-based solution, presents software vulnerabilities that attackers can exploit to break into it and take partial or complete control of its virtual environment. This is indeed a main counterpart with respect to legacy systems not based on software solutions.

Vulnerabilities in the C-RAN scenario can be brought about mainly by two types of software routine that are continuously executed and updated: (i) software patches and (ii) virtual software monitors. The former, software patches, are pieces of software that update the functionalities and/or correct mistakes in computer programs, and they are the current *de facto* solution for the maintenance of software products. The latter, virtual software monitors, are used to supervise the state of virtual systems (such as virtual BBUs, routers, databases, etc.). The main objective of implementing monitors relies on the possibility of validating inner processes and detecting problems. These monitors can inspect and interpret data stored in the network infrastructure, which enables the software monitors to observe the virtual systems installed at any instant [9]. The requirement to apply patches and to control the virtual processes remotely can be considered the main entry point for intrusion attacks, since the corruption or malfunction of any piece of code can jeopardize the system permanently. An attacker can infect a software routine before it is delivered to the network infrastructure provider, and then wait until it is executed to perform the intrusion attack.

9.3.1.2 Uncontrolled Hypervision and Introspection

Introspection is a common capability of virtual systems, in which control software applications, that is monitors, have access to the data stored in the network infrastructure. Although monitors are of paramount importance to oversee the proper functioning of the virtual environment, an attacker that takes control of one of these monitors gets pride of place to eavesdrop sensitive data stored in the virtual BBU pool [10].

A misconfiguration of the access policies applied to software monitors can worsen the situation if the monitors are enabled to copy or modify data. The ability for monitors to copy or duplicate virtual instances for inspection is, indeed, commonly enabled in virtual environments. For example, if the status of a database or a virtual router has to be inspected in a specific moment or time window, the network infrastructure should be configured to enable the monitor to take snapshots of the virtual instances, hence allowing duplication. These copies could be handled and passed through different software applications on demand, and even transferred remotely [11], which leverages the potential of eavesdropping attacks.

Beyond the simple inspection or illicit modification of virtual instances, corrupted software can take advantage of network infrastructure misconfigurations to execute commands in the host operating system. This issue has been reported in previous virtual machine frameworks [12–15]. Once a corrupted software routine manages to execute commands in the host operating system, it has the potential to create rogue instances, modify licit instances of virtual systems such as virtual BBUs, virtual routers and databases, or execute software routines in the host operating system.

9.3.1.3   Rogue Recovery Procedures

Backup databases used for recovery procedures can provide another entry point for intruders, since a corrupted software routine could modify the system backup, or the backups of virtual BBUs or routers, in order to control these entities when they are initiated after a recovery procedure. With this mechanism, an attacker can infect a database and remain undetected until a rollback operation is executed and the rogue entities initiated. Attackers can gain access to the recovery data in shared database environments, where data storage resources are shared by all virtual entities, which is normally the case. It is important for databases to be protected with well-defined access policies, since attackers can take advantage of misconfigured access rules or privilege escalation policies to access and modify the database backup section. Finally, it is worth noticing that an attacker can force an outage of the system through a denial of service (DoS) attack, which will lead the system to roll back and load a corrupted version of the virtual instances, hence enabling the attacker to take control of the system.

9.3.2   Technical Challenges for Intrusion Detection Counter-mechanisms

In the previous section, we detailed different means for an intruder to get into the system by exploiting software vulnerabilities or misconfigurations in the network infrastructure of the virtual entities, and also how an intruder that is inside the system can spread by initiating or modifying other virtual resources inside the host system. It is therefore relevant to discuss the technical difficulties in detecting or avoiding malfunctioning/malicious entities from spreading and affecting other licit virtual instances. In this regard, the main countermeasures to reduce the activity scope of malicious entities inside a virtual environment are based on execution isolation and virtual firewalls. However, these mechanisms present technical challenges arising from the nature of virtual environments.

9.3.2.1 Firewall and Access Control Vulnerabilities

A technical difficulty in protecting C-RANs from intruders is the unavailability of the common security solutions and practices used in physical environments to protect virtual environments. The main reason for this unavailability is that the data-transfer mechanism follows the host operating system socket architecture and not a conventional IP network. This hinders, for example, the application of intrusion detection based on traffic inspection and filtering. Although there are already mechanisms available to inspect virtual entities' communications in the host system backplane, these mechanisms are much more complex. This is due to the heterogeneity in the communication architectures of different host systems and the dynamic nature of virtual environments, characterized for a continuous creation and migration of virtual entities.

Firewall mechanisms are essential in virtual environments, since a rogue virtual entity (virtual router, virtual BBU or software routine) can obtain information inside the virtual environment and transmit this information to the outside world, out of the scope of the C-RAN security mechanisms, or communicate in the backplane with other rogue virtual entities to coordinate and perform collusion attacks. Communication restriction through firewall rules [16] or subnetting (i.e., isolation of virtual entities between different realms) [17] is commonly suggested to prevent information leakage or collusion attacks. However, the firewall policies must be located in the host system backplane and not in the virtual entity. This is because placing the firewall mechanism inside the virtual entity would allow rogue virtual instances to subvert firewall policies, and this would require an in-depth introspection to be detected. Moreover, access policies must also be deployed to avoid modification or eavesdropping of other virtual entities' activities and data. It is worth mentioning that virtual environments deploy shared data-storage resources that can be a soft target for eavesdropping if the access policies are misconfigured and do not work properly.

9.3.2.2 Subversion of Datalogs and Shared Databases

Data storage is another resource commonly shared in virtual environments, since the initiation of dedicated databases per virtual entity would be inefficient. Database sharing allows virtual entities to include and extract data from a common data store, but the access must be controlled through the definition of access policies. This leads to the challenge described above of defining access policies in a dynamic environment where virtual entities appear and disappear constantly, and it could eventually enable rogue virtual entities to eavesdrop or modify the information of other entities. This issue becomes even more relevant when the database maintains the systems' datalogs, since a malicious user could subvert the information stored in the datalogs to remove the entries referring to its illicit actions, hence hindering security forensics. Datalog subversion could also happen by misconfigurations in the host system, such as uncontrolled roll-back procedures, where databases get overridden with a previous state, hence deleting entries in the datalog that describe actions happening after the loaded state [10].

Therefore, security forensics are more difficult to apply in the virtual environment due to the possibility of subverting or losing information, which hinders the traceability of attackers. Moreover, the datalogging task is, itself, more complex in the virtual environment since entities being logged are temporarily available, due to migration or multiple inititation. For instance, a virtual entity that misbehaves at a given time can be replicated or disappear and be initiated in a later period with different identifiers, making it harder to identify the entity responsible for a given illicit action and establish responsibilities.

9.3.3   Insider Attacks

Insider attacks can come from different sources inside the virtual environment, depending on where the rogue entity resides. The most harmful attacks stem from the initiation of rogue virtual routers and virtual BBUs, since these entities perform the core tasks of the C-RAN. Attacks can also be mounted in collusion with rogue users out of the C-RAN that perform remote attacks covered by malicious inside entities. In this section, we differentiate among user-side attacks, triggered by mobile users out of the C-RAN but assisted by intruders, rogue virtual BBU attacks and rogue virtual router attacks.

9.3.3.1   User-side Attacks

Rogue users can perform a series of attacks that are more effective and harder to detect when they are covered by a rogue virtual BBU or assisted by a virtual router. For instance, user identity fraud is feasible in collusion with a rogue virtual BBU or router and remote DoS attacks are more effective in collusion with an intruder.

Regarding identity fraud, a simple ARP table poisoning by a rogue virtual router can direct the packets intended for a licit user towards a rogue user. ARP poisoning [17, 18] is a rather simple-to-perform impersonation, and it only requires a rogue virtual router to access the ARP table and modify its entries, making the impersonation attack undetectable. Identity fraud can also be accomplished with the collusion of a rogue virtual BBU that renews the identities of rogue users by applying user-removal and user-addition procedures [19]. Rogue users can make use of their new identities to inject packets for a DoS attack and deceive others about the source of those packets [20].

Remote DoS attacks can also be performed in collusion with rogue virtual routers, which export configuration files and protocol binaries to rogue users [21]. The users can then inject data packets that exploit the vulnerabilities of virtual routers to break the service down. It is worth noticing that virtual routers are based on programmable packet processors, which present software vulnerabilities that can be exploited by malicious users. A user can inject data packets that modify the operation of the router, leading to a DoS attack [21]. Moreover, these attacks can be masked with identity fraud techniques, making the attacker harder to track.

The users can also obtain the virtual router's state and configuration from an intruder performing a man-in-the-middle attack during a migration procedure. Virtual environments are subject to continuous migration and initiation of virtual entities. The migration of virtual routers and virtual BBUs involves the transmission of configuration files and protocol binaries [22, 23], enabling intruders to obtain sensitive and detailed information merely by sniffing out the migration-related communication. Then the intruder can transmit this information to outside users, easing DoS attacks.

9.3.3.2   Rogue Virtual BBU Attacks

As they are at the core of the C-RAN architecture, virtual BBUs have the potential to perform effective DoS and eavesdropping attacks [21]. A rogue virtual BBU can flood the network infrastructure with resource requests to deplete the resources and hence affect other virtual BBUs. Even if the host system denies the resource requests of the rogue virtual BBU, the intruder can flood the control backplane with requests to lower the performance of the resource-assignment mechanism in the host system.

The privileged position of the virtual BBU inside the virtual environment, connected to the virtual routers, databases and other virtual BBUs, allows it to perform effective side-channel attacks, allowing the attacker to modify or sniff out other BBUs' activities. A rogue virtual BBU can eavesdrop and replicate another virtual BBU, including its traffic. This allows, for example, a rogue virtual BBU to intercept and replicate online services such as proprietary video streaming and redirect them to unauthorized users [21]. In this case, the rogue virtual BBU does not intend to provoke a service disruption but to access the private content of other users, hence the intrusion is harder to detect than in a DoS attack. This kind of side-channel attack has already been described for Amazon cloud services in [24].

A rogue virtual BBU also has the potential to disrupt user services selectively by injecting protocol-specific packets. An example of this kind of attack is described in [25], where a service provider blocked P2P services by sniffing out the packets and injecting forged RESET packets. This simple action broke down Gnutella and BitTorrent services.

9.3.3.3   Rogue Virtual Router Attacks

A rogue virtual router can modify the data transmission flow in several ways:

- Modifying the destination address;
- Intercepting and creating forged packets;
- Inspecting packet content;
- Dropping packets.

Modifying the destination address is not performed directly by modifying the packet, since this would invalidate the integrity checks, but by poisoning the ARP table, as described in Section 9.3.3.1. The virtual router can select the destination node to impersonate and transmit all his intended packets to a rogue user only by modifying one entry in the ARP table. This table is shared by the virtual routers inside the same virtual environment, hence making this attack highly efficient. ARP protection mechanisms are therefore fundamental.

Intercepting and creating forged packets has also been described in Section 9.3.3.2, where the case of creating RESET packets to break down P2P services by Comcast Corporation (one of the biggest Internet service providers in the US) was referred to. It is worth mentioning that this attack can be performed from the virtual routing level, since the routers are composed of reprogrammable packet processors, hence their functionality can easily be modified to create and inject packets and not only to process receiving packets like conventional routers.

Packet inspection in the router level is also a threat that can allow a rogue virtual router to profile users. It is worth noticing that even in encrypted transmissions, traffic analysis can be effective [26] by analysing the identities of the communication parties, the packet size, frequency of transmissions, timetables, etc. The danger of an attacker getting this information goes beyond the simple eavesdropping of users' activities, since it gives attackers valuable information to detect vulnerable source or destination points within the network [27]. For example, with traffic analysis, an attacker can infer which entities control the most accessed online services and launch a DoS attack on those entities, hence increasing the damage.

The simplest attack performed by a virtual router consists of a selective packet drop in the queue of the virtual router. This is a simple and effective way to induce congestion and force a source node to decrease the transmission rate, hence leaving more bandwidth for other source nodes. This attack is especially harmful due to its simplicity and efficiency, since it is difficult for the source or the receiver nodes to elucidate if the decrease in bandwidth is due to congestion or has been induced by router misbehaviour.

9.4 Distributed Denial of Service (DDoS) Attacks Against C-RAN

DDoS attacks are potential availability threats for the C-RAN architecture that can cause severe damage to the network service, since a C-RAN provides service to thousands of users. DDoS attacks are a variation on the typical denial of service (DoS) attacks, where a single attacking host targets a single victim. DoS attacks are one of the oldest and most serious threats on the Internet. Their main objective is to prevent legitimate access to services of a target machine by overwhelming its resources (e.g., CPU, memory, network bandwidth).

There are two main categories of DoS attacks: network-layer DoS attacks and application-layer DoS attacks. Network-layer DoS attacks are carried out at the network layer and they attempt to overwhelm the network resources of the targeted victim with bandwidth-consuming assaults such as TCP SYN or UDP flooding attacks. By contrast, application-layer DoS attacks are more sophisticated attacks that exploit specific characteristics and vulnerabilities of application-layer protocols (e.g., http, DNS or SMTP) and applications running on the victim's system in order to deplete its resources [28–31]. In contrast to DoS attacks, DDoS attacks make use of botnets (i.e., traditional or mobile) in order to deploy multiple attacking entities, often located in disparate locations, and achieve their goal. A traditional botnet is a network of compromised machines (e.g., legitimate PCs, laptops), commonly referred to as bots, which are under the control of an attacker through central command and control (C&C) servers. Hence, the attacker is able to access and manage the traditional botnet remotely via the central C&C servers [30–33]. On the other hand, a mobile botnet is a network of compromised smartphones (i.e., bots) remotely controlled by a bot-master via C&C channels [32, 34].

In the upcoming 5G communication systems, mobile botnets are expected to be increasingly used by attackers, since smartphones are ideal remote-controlled machines due to their specific features. They support different connectivity options and increased uplink bandwidth, and tend to be always turned on and connected to the Internet. Consequently, attackers will be enabled to deploy mobile botnets in many efficient ways and launch severe attacks against other legitimate mobile users or against components of the mobile communication system (e.g. the access network, the mobile operator's core network) [34–36]. In this sense, potential DDoS attacks against C-RAN can be launched from mobile botnets, as shown in Figure 9.3. The bot-master, which is the malicious actor, will be responsible for choosing the mobile devices that will be compromised by malware and turned into bots. Specifically, the bot-master will exploit security vulnerabilities (e.g., operating system and configuration vulnerabilities) of the chosen mobile devices and compromise them. In current mobile botnets, the bot-masters use similar http techniques to those used by the PC-based botnets and new techniques specific to mobile devices' features such as SMS messages, in order to distribute their commands. Furthermore, the bot-proxy servers will be the means of communication that the bot-master will use to command and control the bots indirectly. Finally, the bots will be programmed and instructed by the bot-master to perform DDoS attacks against C-RAN [34–36]. Examples of potential DDoS attacks against C-RAN are described in detail below.

9.4.1 DDoS Attacks Using Signalling Amplification

Signalling amplification attack has been described for 4G networks in [37], and this attack can be extended to the C-RAN network architecture. As depicted in Figure 9.3, this attack can be triggered by a network of infected mobile devices (i.e. a mobile

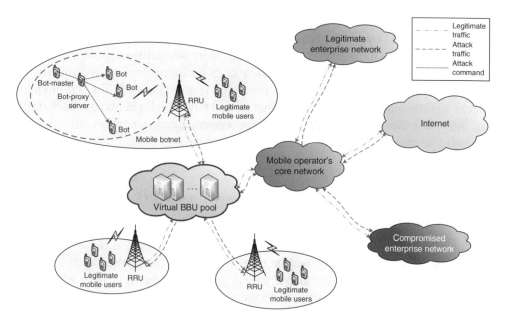

Figure 9.3 C-RAN architecture under potential attacks in a 5G communication scenario

botnet) under the same C-RAN, but can be located at one or more of the RRUs covered by the C-RAN. The objective of the signalling amplification attack is to overload the resource management unit by flooding the C-RAN with signalling overhead. In particular, the mobile botnet transmits signalling used to establish and release dedicated radio bearers. The concurrent transmission of this signalling overhead causes the virtual BBUs in charge of resource management tasks to initiate the bearer activation and assignment process. It is worth noticing that this attack can be performed from multiple RRUs simultaneously, and the resource requests will be managed by several virtual BBUs hosted in the same C-RAN, hence depleting the processing resources. Moreover, after the bearers have been assigned to the bots, they will not use them and the expiration of the assignation, which will be triggered by a timeout, will lead the virtual BBUs to proceed with the bearer deactivation procedure, which also contributes to depleting C-RAN resources. The botnet can be programmed to perform this attack repeatedly. The detection mechanism, proposed in [37], for this kind of attack in LTE networks is based on tracking the inter-setup time and the number of bearer activations/ deactivations per minute, and comparing these values with predefined thresholds.

9.4.2 DDoS Attacks Against External Entities Over the Mobile Network

DDoS attacks against targets in external networks (e.g., enterprise networks) connected to the mobile core network are possible to carry out over the upcoming mobile communication systems. In this scenario, a botnet of mobile devices can be used to generate a high volume of traffic against the victim, located in a legitimate enterprise network, over the mobile operator's core network, as shown in Figure 9.3. Although the target of these attacks will not be the mobile network itself, the fact that they inject large traffic loads into it can impact its performance, and specifically, the performance of the C-RAN that serves the traffic from the botnet. The recent DDoS attacks against Spamhaus over the Internet proved how a high volume of attack traffic can affect the availability of the underlying communication network employed to transmit it to the specific target [38].

9.4.3   DDoS Attacks from External Compromised IP Networks Over the Mobile Network

The upcoming mobile communication systems may not only suffer from DDoS attacks originated by malicious mobile users (i.e. mobile botnets) covered by RRUs associated with the C-RAN, but also from compromised external IP networks (e.g., a compromised enterprise network) connected to the mobile core network, as depicted in Figure 9.3. These networks will encompass many compromised devices/machines that will generate a considerable volume of traffic against an entity being served by the C-RAN or against the C-RAN itself. The key issue in this scenario is that the devices/machines included in external IP networks, connected to the mobile core network, can be a soft target for being compromised by malware, through infected mobile devices accessing these networks and, thus, creating traditional botnets (e.g., PC-based) or mobile botnets to launch effective DDoS attacks from these networks.

   For example, malware infection of enterprise network devices/machines can be caused by infected employees' devices [39]. It is a common trend for employees to bring their own smartphones to the working environment and connect them to enterprise networks or even to enterprise networks with strict access control. However, smartphones are susceptible to mobile malware and thus, attackers will make use of the appropriate malware that will allow them to compromise an otherwise secure enterprise network and then infect its devices/machines [39]. It is worth commenting that the multiple connectivity technologies, such as 2G/3G/4G/5G, WiFi, Bluetooth, NFC (near field communication) and USB (universal serial bus), supported by smart-phones can be abused by attackers as mobile malware propagation channels. Hence, employees' smartphones can work as bridges for attackers between the enterprise network and the outside world. As a result, an employee's smartphone can be com-promised through a mobile communication channel (e.g., 3G/4G/5G) or a short-range communication channel (e.g., NFC) and become a wormhole to the target enterprise network or bring a malicious payload directly to it through another communication channel (e.g., USB) supported by the smartphone [34, 39]. To avoid security breaches of the enterprise network arising from the use of employees' smartphones inside the working environment, a very common approach is to periodically scan all employees' smartphones with anti-malware software. Nevertheless, this approach is intrusive and too costly energy-wise. Consequently, innovative solutions providing a balance between security responsiveness and cost-effectiveness are required. In [39], strategic sampling has been proposed as a solution to address this requirement by identifying and periodically sampling security representative smartphones. The sampled smartphones are then checked for malware infections.

9.5 Conclusions

C-RAN technology has to deal with security issues which are absent in the previous RAN technology and are associated with the adoption of cloud computing and virtual systems, that is, virtualized services and networks, for the implementation of RAN services. These security issues can lead to very destructive results due to the single-point deployment strategy foreseen for the C-RAN architecture. In this chapter, we have presented potential threats and attacks against the main components of the C-RAN architecture, which stem from virtualizing networks and processing units in a single hardware point. Malicious users can take advantage of the C-RAN single-point deployment strategy to disrupt or take control of the C-RAN services by attacking a single entity. In particular, we have focused on examples of potential intrusion attacks and DDoS attacks. The potential intrusion attacks represent very challenging attacks, requiring the attacker to take control of a virtual entity inside the C-RAN virtual environment, which can be achieved by exploiting vulnerabilities such as misconfigurations or infected software routines. An intruder could initiate rogue virtual entities, opening a range of possibilities for multiple attacks such as illegal introspection of licit virtual entities, private data eavesdropping, data or service replication, user impersonation and service disruption. On the other hand, the possible DDoS attacks against the C-RAN architecture can be launched remotely by botnets and cause severe damage to the network service.

References

[1] Wang, C.-X., Haider, F., Gao, X., You, X.-H., Yang, Y., Yuan, D., Aggoune, H., Haas, H., Fletcher, S. and Hepsaydir, E. (2014) Cellular architecture and key technologies for 5G wireless communication networks. *IEEE Communications Magazine*, **52**(2), 122–130.

[2] Chih-Lin, I., Rowell, C., Han, S., Xu, Z., Li, G. and Pan, Z. (2014) Toward green and soft: a 5G perspective. *IEEE Communications Magazine*, **52**(2), 66–73.

[3] Bangerter, B., Talwar, S., Arefi, R. and Stewart, K. (2014) Networks and devices for the 5G era. *IEEE Communications Magazine*, **52**(2), 90–96.

[4] Checko, A., Christiansen, H. L., Yan, Y., Scolari, L., Kardaras, G., Berger, M. S. and Ditmann, L. (2014) Cloud RAN for Mobile Networks – a Technology Overview. *IEEE Communications Surveys and Tutorials*, **17**(1), 1.

[5] Wang, R., Hu, H. and Yang, X. (2014) Potentials and Challenges of C-RAN Supporting Multi-RATs Toward 5G Mobile Networks. *IEEE Access*, **2**, 1187–1195.

[6] Panwar, N., Sharma, S. and Kumar Singh, A. (2015) *A survey on 5G: The next generation of mobile communication, Physical Communication*. Available online as at 11 November 2015, ISSN 1874–4907.

[7] Wu, J., Zhang, Z., Hong, Y. and Wen, Y. (2015) Cloud radio access network (C-RAN): A primer. *IEEE Network*, **29**(1), 35–41.

[8] Holvitie, J., Leppanen, V. and Hyrynsalmi, S. (2014) Technical Debt and the Effect of Agile Software Development Practices on It – An Industry Practitioner Survey. In the *Sixth International Workshop on Managing Technical Debt (MTD)*, pp. 35–42, September.

[9] Whitaker, A., Cox, R. S., Shaw, M. and Gribble, S. D. (2005) Rethinking the design of virtual machine monitors. *Computer*, **38**(5), 57–62.

[10] van Cleeff, A., Pieters, W. and Wieringa, R. J. (2009) Security implications of virtualization: A literature study. In *Proceedings of the International Conference on Computational Science and Engineering*, IEEE Computer Society, Washington, DC, USA.

[11] Roschke, S., Cheng, F. and Meinel, C. (2009) Intrusion detection in the cloud. In *Proceedings of the IEEE International Conference on Dependable, Autonomic and Secure Computing*, IEEE Computer Society, Washington, DC, USA.

[12] Common Vulnerabilities and Exposures (2012) CVE-2012-1516 https://cve.mitre.org/cgi-bin/cvename.cgi?name=CVE-2012-1516.

[13] Common Vulnerabilities and Exposures (2012) CVE-2012-1517. https://cve.mitre.org/cgi-bin/cvename.cgi?name=CVE-2012-1517.

[14] Common Vulnerabilities and Exposures (2012) CVE-2012-2449. https://cve.mitre.org/cgi-bin/cvename.cgi?name=CVE-2012-2449.

[15] Common Vulnerabilities and Exposures (2012) CVE-2012-2450. https://cve.mitre.org/cgi-bin/cvename.cgi?name=CVE-2012-2450.

[16] Wolinsky, D. I., Agrawal, A., Boykin, P. O., Davis, J. R., Ganguly, A., Paramygin, V., Sheng, Y. P. and Figueiredo, R. J. (2006) On the design of virtual machine sandboxes for distributed computing in wide-area overlays of virtual workstations. *International Workshop on Virtualization Technology in Distributed Computing*, IEEE Computer Society, Washington, DC, USA.

[17] Wu, H., Ding, Y., Winer, C. and Yao, L. (2010) Network security for virtual machine in cloud computing. In *Proceedings of the 5th International Conference on Computer Sciences and Convergence Information Technology (ICCIT)*, Seoul, South Korea.

[18] Cavalcanti, E., Assis, L., Gaudencio, M., Cirne, W. and Brasileiro, F. (2006) Sandboxing for a free-to-join grid with support for secure site-wide storage area. In *Proceedings of the International Workshop on Virtualization Technology in Distributed Computing*, IEEE Computer Society, Washington, USA.

[19] Chowdhury, N. M. M. K., Zaheer, F.-E. and Boutaba, R. (2009) imark: An identity management framework for network virtualization environment. In *Proceedings of the IFIP/IEEE International Symposium on Integrated Network Management*, IEEE Press, Piscataway, USA.

[20] Cabuk, S., Dalton, C. I., Ramasamy, H. and Schunter, M. (2007) Towards automated provisioning of secure virtualized networks. In *Proceedings of the ACM Conference on Computer and Communications Security*, New York, USA.

[21] Natarajan, S. and Wolf, T. (2012) Security issues in network virtualization for the future Internet. In *Proceedings of the International Conference on Computing, Networking and Communications (ICNC)*, January 30–February 2, pp. 537–543.

[22] Wang, Y., Keller, E., Biskeborn, B., van der Merwe, J. and Rexford, J. (2008) Virtual routers on the move: Live router migration as a network-management primitive. *SIGCOMM Computer Communication Review*, **38**, 231–242.

[23] Clark, C., Fraser, K., Hand, S., Hansen, J. G., Jul, E., Limpach, C., Pratt, I. and Warfield, A. (2005) Live migration of virtual machines. In *Proceedings of the 2nd Conference and Symposium on Networked Systems Design and Implementation (NSDI)*, Berkeley, California, **2** pp. 273–286.

[24] Ristenpart, T., Tromer, E., Shacham, H. and Savage, S. (2009) Hey, you, get off of my cloud: Exploring information leakage in third-party computer clouds. In *Proceedings of the 16th ACM Conference on Computer and Communications Security (CCS)*, New York, pp. 199–212.

[25] FCC rules against BitTorrent blocking. EFF. Available at: https://www.eff.org/es/deeplinks/2008/08/fcc-rules-against-comcast-bit-torrent-blocking

[26] Conti, M., Mancini, L. V., Spolaor, R. and Verde, N. V. (2016) Analyzing Android Encrypted Network Traffic to Identify User Actions. *IEEE Transactions on Information Forensics and Security*, **11**(1), 114–125.

[27] Bays, L. R., Oliveira, R. R., Barcellos, M. P., Gaspary, L. P. and Madeira, E. R. M. (2015) *Virtual Network Security: Threats, Countermeasures, and Challenges*. Springer, London.

[28] McGregory, S. (2013) Preparing for the next DDoS attack. *Network Security*, **5**, 5–6.

[29] Zargar, S. T., Joshi, J. and Tipper, D. (2013) A survey of defense mechanisms against distributed denial of service (DDoS) flooding attacks. *IEEE Communications Surveys and Tutorials*, **15**(4), 2046–2069.

[30] Mantas, G., Stakhanova, N., Gonzalez, H., Jazi, H. H. and Ghorbani, A. A. (2015) Application-layer denial of service attacks: taxonomy and survey. *International Journal of Information and Computer Security*, **7**(2–4), 216–239.

[31] Logota, E., Mantas, G., Rodriguez, J. and Marques, H. (2015) Analysis of the Impact of Denial of Service Attacks on Centralized Control in Smart Cities. In *Mumtaz, S., Rodriguez, J., Katz, M., Wang, C. and Nascimento, A. (eds) Wireless Internet*, Springer International Publishing, pp. 91–96.

[32] Hoque, N., Bhattacharyya, D. K. and Kalita, J. K. (2015) Botnet in DDoS Attacks: Trends and Challenges. *IEEE Communications Surveys and Tutorials*, **17**(4), 2242–2270.

[33] Freiling, F. C., Holz, T. and Wicherski, G. (2005) *Botnet Tracking: Exploring a root-cause methodology to prevent distributed denial-of-service attacks*. Springer, Berlin, Heidelberg.

[34] Mantas, G., Komninos, N., Rodriguez, J., Logota, E. and Marques, H. (2015) *Security for 5G Communications: Fundamentals of 5G Mobile Networks*, John Wiley & Sons.

[35] Arabo, A. and Pranggono, B. (2013) Mobile Malware and Smart Device Security: Trends, Challenges and Solutions. In *Proceedings of the 19th International Conference on Control Systems and Computer Science (CSCS)*, pp. 526–531.

[36] Flo, A. R. and Josang, A. (2009) Consequences of botnets spreading to mobile devices. In *Short-Paper Proceedings of the 14th Nordic Conference on Secure IT Systems (NordSec)*, pp. 37–43.

[37] Bassil, R., Chehab, A., Elhajj, I. and Kayssi, A. (2012) Signaling oriented denial of service on LTE networks. In *Proceedings of the 10th ACM International Symposium on Mobility Management and Wireless Access (MobiWac)*, New York, pp. 153–158.

[38] Piqueras Jover, R. (2013) Security attacks against the availability of LTE mobility networks: Overview and research directions. In *Proceedings of the 16th International Symposium on Wireless Personal Multimedia Communications (WPMC)*, pp. 1–9, June.

[39] Li, F., Peng, W., Huang, C.-T. and Zou, X. (2013) Smartphone strategic sampling in defending enterprise network security. In *IEEE International Conference on Communications (ICC)*, pp. 2155–2159, June.

Index

Backhauling/Fronthauling for Future Wireless Systems, First Edition.
Edited by Kazi Mohammed Saidul Huq and Jonathan Rodriguez.
© 2017 John Wiley & Sons, Ltd. Published 2017 by John Wiley & Sons, Ltd.